The Basics
of
Technical
Communicating

The Basics
of
Technical
Communicating

MYLAR
SEP
CHEM

B. Edward Cain

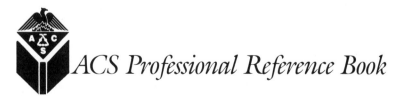

ACS Professional Reference Book

American Chemical Society
Washington, DC 1988

Library of Congress Cataloging-in-Publication Data

Cain, B. Edward, 1942–
   The basics of technical communicating.

"ACS professional reference book."

Includes bibliographies and index.

   1. Communication of technical information.
2. Communication in chemistry.   I. American Chemical
Society.   II. Title.

T10.5.C25   1988   808'.0665   88–3325

ISBN 0–8412–1451–4
ISBN 0–8412–1452–2 (pbk.)

| About the Author |

B. Edward Cain received a B.A. degree from Harpur College, State University of New York (SUNY) at Binghamton, in 1964 and earned a doctorate in inorganic chemistry at Syracuse University in 1971. After holding positions as a chemist for a public health laboratory and a consulting engineering firm, he joined the faculty of the National Technical Institute for the Deaf (NTID), a college of Rochester Institute of Technology (RIT). He later transferred to the Department of Chemistry, where he is currently a full professor.

Dr. Cain's interest in technical communication began when RIT adopted a Writing Proficiency Policy for undergraduates. Since that time, he has been highly involved in implementing the requirement for chemistry majors. His other professional efforts are adapting chemistry lectures and labs for physically handicapped students, curriculum development, recruitment, and teaching chemistry to nonscience majors. He has written several papers and is a recipient of the Eisenhart Annual Award for Outstanding Teaching at RIT.

Dr. Cain is a member of the American Chemical Society (ACS), the American Association for the Advancement of Science (AAAS), the American Association of University Professors (AAUP), the National Association of the Deaf (NAD), the National Registry of Interpreters for the Deaf (NRID), and the National Science Teachers Association (NSTA). He has served on advisory committees addressing science and the handicapped for the ACS, AAAS, and NSTA, and he has held office as member-at-large in the Rochester section of the ACS. Dr. Cain's profile is given in *Who's Who of Emerging Leaders*, *Who's Who in the East*, and *Personalities in America*.

| Contents |

Index

| Preface |

A recent newspaper headline, "Students Flunking as Writers", described a survey issued by the National Assessment of Educational Progress (NAEP). NAEP evaluated the writing abilities of 95,000 17-year-olds, and their survey showed that 76 percent of the students could not write an adequate imaginative piece of writing, 80 percent could not write a persuasive letter, and 62 percent wrote unsatisfactory informative prose (*1*).

According to the article, these results have not changed significantly throughout the past 15 years despite the efforts of educators at all levels. Yet there is an increasing awareness that students who are science and technology majors—who will become professionals in science and engineering—need to have technical writing and oral skills *as part of* the basic tools required for their disciplines. Technical ability in a scientific discipline is simply not enough for a person whose goal is to achieve a higher level of professional performance. A direct relationship exists between the ability to communicate effectively and the opportunity to advance.

The American Chemical Society and other professional organizations have recognized the need to emphasize the skill of communicating, and a growing number of publications and symposia have addressed the topic "Writing Across the Curriculum". Improving communication skills is not limited to a student who is in college, however; it is a continuing process and can be more difficult for a person who is already employed. This book is intended not only for students who are science and technology majors but also for on-the-job professionals who wish to improve their skills.

Why is there a need for yet another book about this subject? If you look in a library or bookstore for books on technical writing or communication in science, you will find a large selection. However, most of these books are not tailored to the special needs of chemists and other professional scientists.

Everyday activities require professional scientists to be able to communicate well. Memos have become commonplace (too commonplace?) in the business and academic worlds. Technologies are changing so quickly that accurate documentation is continually required. Because litigation is always a potential threat, we must make sure that our records are technically and grammatically correct in expressing our actions. In addition, routine documents such as proposals, bids, specifications, personnel evaluations, requisitions, and instructions must also be written with expertise.

As members of the scientific community, we have all had to learn a variety of languages and their applications: foreign languages, computer languages, and the specific languages of the various disciplines (such as chemistry, biology, physics, engineering, and mathematics). It is increasingly important to be able to communicate the content of these technical languages to one another and to the public. Scientists and technologists must be able to explain phenomena such as Chernobyl, AIDS, and genetic engineering (which are covered by the mass media) to a somewhat scientifically illiterate audience.

It is not my intention to teach grammar, spelling, or other parts of basic English; many excellent resources are available for that purpose. My goal is to demonstrate a variety of applications of using correct English to communicate effectively. Because I am a chemist, many of the examples given will be from the chemistry profession. However, with slight changes, the information should be applicable to scientists and engineers in all disciplines. Formats may have to be altered to fit specific needs, but the basics of correct technical communication vary only slightly.

This volume will complement *The ACS Style Guide* (2) and *Writing the Laboratory Notebook* (3). All three books should provide a complete package for professional development in technical communication skills.

Acknowledgments

I am especially grateful for the assistance provided by the ACS Books Department. The Rochester Institute of Technology was generous in granting me a full quarter of leave to prepare this manuscript. Without the encouragement and patience of colleagues and friends, this task would not have been undertaken and completed.

References

1. *Democrat and Chronicle*; Rochester, NY; April 13, 1986; p 19A.
2. *The ACS Style Guide*; Dodd, J. S., Ed.; American Chemical Society: Washington, DC, 1986.
3. Kanare, H. M. *Writing the Laboratory Notebook*; American Chemical Society: Washington, DC, 1985.

B. EDWARD CAIN
Rochester Institute of Technology
Rochester, NY 14623–0887

July 31, 1987

IMPROVING YOUR TECHNICAL COMMUNICATION SKILLS

| Chapter 1 |

What Is Technical Communication?

If asked to define technical communication, most people might respond that it is the ability to express information about technical areas, such as chemistry, engineering, or computer science, in technical terms. That is exactly what most people believe, but it is not a complete definition.

Let us first define the nontechnical parts of communicating because these are the areas in which most people have experience. In school, we have all been exposed to creative writing: poems, short stories, and novels. We have also had to express ourselves in expository writing: essays, editorials, letters to the editor, and debates. These opportunities have allowed us to express opinions and often to try to convince others that our opinions are correct. These opinions did not have to be factually based and could stem from emotions, religious beliefs, or "gut-level" feelings.

Technical communication is unlike either creative or expository expression in that it must be based on fact. It requires clear, precise, unambiguous, unemotional, and sometimes nonjudgmental language. However, it does not need to be dull. Opinions can be expressed, but they must be substantiated by facts. Persuasion can be exercised, but it must have a factual rather than an emotional basis.

Technical communication is always in response to a need. Someone has asked for the information or you feel that there is a need to

1451–4/88/0003$06.00/1 © 1988 American Chemical Society

provide it. Perhaps the need is for permission or funding:

- a proposal to follow a specific line of research or change its direction
- a proposal to obtain grant money to do research
- a proposal for a piece of equipment necessary to execute a specific function
- a bid for permission to provide services to a company

The need might be to provide information, either requested or volunteered:

- a journal article describing research results
- a weekly progress report requested by your supervisor
- a cost estimate for a specified project or task
- an oral report to the Board of Directors on a new product line

The need might be to answer a question or to provide alternatives:

- The company may want a comparative study of different manufacturers' equipment to make a decision about a purchase.
- A prospective employer might want to know why you think you are qualified for the position.
- A company might want you to provide alternatives to their existing process.
- Your supervisor might ask why the project is unfinished.

The need might be to provide instructions:

- a journal article describing a new method or application
- a manual documenting the use of new equipment
- directions for using software

In short, technical communication can be and must be used in a variety of ways. Most often, the categories overlap; for example, a document to be used in decision making would need to answer questions, provide information, solve problems, and address a specific task.

Requirements for Technical Communication

The most basic requirement for technical communication is that the person providing the information completely understands the subject. This assumption seems elementary, but it is not always the case in the real world.

Fully understanding the topic does not mean knowing every facet of it. For example, a person who writes a computer program to analyze crystallographic data would not have to know all the details of the electronics of the computer; however, he or she would need to be completely familiar with the computer language used, the acquisition and handling of the raw data, and the transformation of the data into the final results. Even then, there can be gaps in a person's background, and the audience should be made aware of those gaps.

Unlike expository writing, technical information should place emphasis on the subject, not on the writer. The writer can state opinions, but they must be based on factual evidence and not solely on personal feelings. Interpretation of results is also not prohibited; indeed, it is often necessary and encouraged. However, the focus should remain on the topic being considered, and the reader needs to know that the statement is a personal opinion or interpretation of the facts.

Technical communication should be tailored to the need, efficient in presentation, and unambiguous. Wordiness can be avoided by focusing on the subject (i.e., not wandering off the track) and by using precise, concise, and effective phrasing to express information. Sentences must be grammatically correct, but the author's style does not need to be lost.

Audience

Because technical communication is always tailored to a specific need, it should be presented with the audience's background in mind. Will one person be the recipient of the information or will many? Will the audience have the same background as the presenter or will it be at a much different technical level? Will the information appear before the general public (e.g., *TIME*, *Newsweek*, or a book intended for a broad audience) or before a specialized group (e.g., *Journal of the American Chemical Society* or *Science*)?

The highly technical report is the easiest type for a trained specialist to present. If the specialist assumes that the audience is at the same level of sophistication in the subject, he or she can ignore the basics and omit long explanations, interpretations, and definitions of specialized vocabulary.

A semitechnical presentation is directed to an audience that has some background in the subject but is not as well-versed as the author. The presenter will need to elaborate on basics before introducing the complexities of the actual topic. Definitions and examples of specialized vocabulary and interpretations of various aspects of the topic should be provided.

The nontechnical report, although the simplest form, is the most difficult for highly trained specialists to write. Although it is impossible to avoid using technical terminology, it should be minimized and extensively defined, with examples. A glossary is often desirable. The report should be clear and to the point. Visual aids are helpful when explaining difficult topics. Good examples of this type of writing can be found in the science sections of *The New York Times, TIME,* or *Newsweek.*

The writer must also consider the secondary audience and its needs, especially with written materials. If the primary audience has a nontechnical background (such as the Board of Directors), it would rely upon advice from experts in the subject. A highly detailed, technical appendix (for the secondary audience, the experts) would need to be attached to the primary, nontechnical report (for the Board). The same type of consideration should be involved in the reverse situation. If the report is created for people knowledgeable in the subject, a nontechnical summary or abstract may be needed for others who might be consulted but would not have the same background.

Summary

In summary, a complete definition of technical communication includes these aspects:

1. It responds to a need.
2. It is based on fact.

3. It requires clear, precise, unambiguous, and unemotional language.
4. It can address a variety of audiences.

This book will guide you to successful approaches for all of these aspects.

| Chapter 2 |

Eliminating Wordiness and Jargon

When you've got a thing to say,
Say it! Don't take half a day . . .
Life is short—a fleeting vapor—
Don't you fill the whole blamed paper
With a tale which, at a pinch,
Could be cornered in an inch!
Boil her down until she simmers,
Polish her until she glimmers.

Joel Chandler Harris (1)

T his advice is as appropriate for scientists today as it was for reporters in the late 1800s. The goals of good technical writing are to eliminate unnecessary words and phrases and to use the most precise words and concise phrases as possible.

Unnecessary Words

Consider the following passage:

The experiment calls for a condenser to be connected **up** to the cold water tap. The solution should be heated **up** so the reaction

1451–4/88/0009$06.00/1 © 1988 American Chemical Society

> will start **up**. This occurs **at** above 77 °C and the directions call
> for heating for one hour until the reaction is over **with**.

This passage might be acceptable in everyday conversation, but it would not be appropriate for proper writing, technical or nontechnical. It contains an excess of prepositions and adverbs that can be deleted to make a less cluttered, more straightforward statement. By removing the words in boldface type, the passage retains its original meaning but looks and sounds better:

> The experiment calls for a condenser to be connected to the cold water tap. The solution should be heated so the reaction will start. This occurs above 77 °C and the directions call for heating for one hour until the reaction is over.

Appendix I lists common phrases that contain unnecessary words.

In addition to extra prepositions and adverbs, our language has many phrases that can and should be replaced by one word. The revised passage could be improved further by changing "calls for" to the more proper meaning "requires" or "specifies":

> The experiment requires a condenser to be connected to the cold water tap. The solution should be heated so the reaction will start. This occurs above 77 °C and the directions specify heating for one hour until the reaction is over.

Appendix II lists common examples of wordy phrases and suggests better words to substitute for them.

Tautology

Another frequent writing error is *tautology*, which is the needless repetition of the same idea using different words in the same sentence. Examples are

> true facts (facts are true)

> equally as good as (if it is as good as, then it's equal)

> initial beginning of the experiment ("initial" is the same as "first" or "the beginning")

Appendix III lists other tautologies that should be avoided.

Redundancy

Redundancy is the use of superfluous words that do not add to the meaning of the statement. Examples are

A domino is rectangular **in shape**.

A domino is black and white **in color**.

An elephant is large **in size**.

In these examples, rectangular is a shape, black and white are colors, and large describes size. The boldface portions of the sentences could be eliminated without interfering with the overall essence of the statements.

Comparative adjectives (e.g., more, most, less, and least) can be used improperly if they modify absolute words. For example, a result cannot be most unique, because unique means "being the only one" or "single in kind". A result can be unique but not most unique.

In everyday language and speech, many absolute words are modified, but only when using a secondary meaning. For example, a person could say, "That party was even more dead than the last one." He or she uses the phrase "even more dead" to mean "less lively" or "less interesting". A living object, however, cannot be "more dead"; it is either dead, alive, or, if in bad health, dying. Appendix IV contains a list of absolute words that should not be modified when used in the sense of their primary definitions.

Imprecise Words

Imprecise or vague words should be avoided in technical writing and replaced by quantitative expressions whenever possible. This practice is especially important when you are describing an event or the odds of the event's occurrence. Consider the following:

only (rarely, seldomly, infrequently, once in a while) did the reaction mixture explode. . .

it is (likely [or unlikely], almost certain) that the product is toxic. . .

the substance was (likely to be, almost always, frequently, usually, generally) contaminated with. . .

the disease (tended to be, was often, was most often) fatal. . .

The words in parentheses are more vague than the reader would wish. How often does the reaction mixture explode? 1 out of 10? 1 out of 1,000,000? The answer to that question could have considerable bearing on the decision to follow the published procedure.

Using imprecise words is not completely prohibited and is often appropriate when describing people or situations that are not predictable. To write that "a person rarely smiled" gives a good mental picture of a typically dour expression, and the qualifier is understood well by the reader. Research in the medical profession (2) indicates an implicit differentiation among doctors as to the meaning of several imprecise phrases; for example, "almost certain" conveys 95% likelihood, and "very likely" indicates about 90% chance.

Superfluous Phrases

Some phrases are simply superfluous, but the writer feels obligated to use them as introductions to various ideas. "Needless to say" is a phrase that causes the reader to respond, "Why say it then?" Several other phrases are in the same category, such as:

As a matter of fact

As you already know

For all intents and purposes

In a manner of speaking

In my opinion

It has come to my attention

It is interesting to note that

It may be said that

It would appear that

Last but not least

The bottom line is

The statement may be made that

Omit these phrases in technical writing; they contribute nothing to the subject and are simply fillers. In an oral presentation, limited use of these phrases is more acceptable. People being bombarded with technical information need some time to assimilate it, and these fillers can provide a short opportunity for the listeners to digest the information!

Overly Complex Words

Writing and reading about technical subjects accustoms people to complex terminology. Unfortunately, using simpler words often becomes difficult or seems trite. "Securing employment" sounds more sophisticated than "finding a job", even though the latter is more clear and to the point. Whenever it is possible to substitute everyday language for a multisyllabic term without resorting to slang or losing precision in the writing, you should do so.

The title of a recent article by Edward Tenner (3) is a good example of excessive use of large words to the extent that the meaning is obscured (purposefully, in this case):

Cognitive Input Device in the Form of a Randomly Accessible Instantaneous-Read-Out Batch-Processed Pigment-Saturated Laminous-Cellulose Hard-Copy Output Matrix*

*Or magazine article in Tech Speak, which replaces the jargon of Indo-European herdsmen with a rigorous and logical language.

Each area of specialization has its own slang or jargon that colleagues use to communicate very effectively among themselves. Acronyms are easier to use on a continued basis than the string of words for which they stand. Governmental agencies, such as the

National Aeronautics and Space Administration (NASA), are notorious for creating new words, but scientists in the fields of chemistry, computer science, and other technologies have all contributed to the constantly changing vocabulary. If the intended audience might include those who are not as informed about the subject as the writer, jargon should be avoided. If it is more efficient to use these terms, they certainly should be defined clearly upon their first use in a paper or talk.

References

1. *The Shorter Bartlett's Familiar Quotations*; Sproul, K., Ed.; Permabooks: New York, 1959; "Advice to Writers for the Daily Press," p 157.
2. Mosteller, F.; as quoted by Kolata, G. *Science* **1986,** 234, 542.
3. Tenner, E. *Discover* **1986,** 7(5), 58–64.

Exercises

1. Prepare a one-page paper about a scientific topic using your natural style of writing (or use a previously written paper). Then carefully reread what you have written, and try to identify phrases that need shortening, omitting, or replacing; tautologies; redundancies; and misuse of comparatives. Rewrite the page with the correct style.

2. Photocopy a journal, magazine, or newspaper article about a scientific subject and identify any of the common errors mentioned in this chapter (maybe there will be none). Rewrite the sentences in which errors occur.

Appendix I. Phrases That Contain Unnecessary Words

Delete the words in boldface type.

as to whether	end **up**	lose **out on**	reduce **down**
at above	enter **in**	meet **up with**	refer **back**
at below	enter **into**	miss **out on**	remain **back**
check **into**	face **up to**	**of** from	repeat **again**
check **on**	file **away**	off **of**	return **again**
check **upon**	follow **after**	open **up**	return **back**

Appendix I.—Continued

climb **up**	follow **up**	out **of**	revert **back**
close **down**	head **up**	outside **of**	rest **up**
close **up**	hoist **up**	over **with**	resume **again**
connect **up**	**in** between	pay **up**	retreat **back**
count **up**	inside **of**	penetrate **into**	seal **off**
debate **about**	join **up**	plan **on**	**still** remain
decide **on**	know **about**	protrude **out**	termed **as**
descend **on**	later **on**	raise **up**	weigh **out**
divide **up**	level **off**	recall **back**	win **out**
empty **out**	link **up**	recur **again**	write **up**

Appendix II. Phrases To Be Replaced by Shorter Words

Phrase	Improved Wording
accounted for by the fact that	due to, caused by, because
a considerable amount of	much
add the point that	add that
afford an opportunity	permit, allow
after this is accomplished	then
a great deal of	much
along the line(s) of	like
a majority of	most
an example of this is the fact that	for example, as an example, thus
another aspect of the situation to be considered	as for
a number of	several, many, some
are of the same opinion	agree
as a consequence of	because
as a matter of fact	in fact
as is the case	as happens
as of now (this date)	now
as per	eliminate; rephrase sentence
as regards	about
as related to	for, about
assuming that	if
as well as	and
at an earlier date	previously
at a rapid rate	rapidly, quickly
at this point in time	now

Appendix II.—Continued

Phrase	Improved Wording
at this time	now
aware of the fact that	know
based on the fact that	due to, because
beg leave to say, beg to differ	eliminate; please don't beg!
by means of	by
come to an end	end
despite the fact that	although
due to the fact that	because
duly noted	noted
during the time that	while
except in a small number of cases	usually
exhibit a tendency to	tend to
first of all	first
for the purpose of	for
for the (simple) reason that	because
give rise to	cause
has the ability to	can
inasmuch as	for, as
in case	if
in close proximity	near, close
in connection with	about, concerning
in consequence of this fact	therefore
in light of the fact that	because
in many cases	often
in order to	to
in regard to	about
in relation to	to, toward
in relation with	with
in respect to	about
in short supply	scarce
in spite of the fact that	although
in terms of	in
in the case of	for, by, in, if
in the course of	during
in the event of (that)	if

Appendix II.—*Continued*

Phrase	Improved Wording
in the first place	first
in the majority of instances	usually
in the matter of	about
in the not-too-distant (near) future	soon
in the proximity of	near, nearly, about
in the vicinity of	near
in this day and age	today
in view of the above	therefore
in view of the fact that	therefore, because, since
it is apparent that (it appears that)	apparently
it is believed that	I believe, I think
it is clear that	clearly
it is doubtful that	possibly
it is incumbent on me	I must
it is often the case that	often
it stands to reason	eliminate if possible
it would not be unreasonable to assume	I assume (or we assume)
kindly send	please send
leaving out of consideration	disregarding
make the acquaintance of	meet
not of a high order of accuracy	inaccurate
notwithstanding the fact that	although
of considerable magnitude	big, large, great
of great importance	important
of very minor importance	unimportant
on account of the fact that	because
on a few occasions	occasionally
on a personal basis	personally
on behalf of	for
on the grounds that	because
on the order of	about

Appendix II.—Continued

Phrase	Improved Wording
owing to the fact that	because
relative to	about
subsequent to	after
take the place of	substitute
the only difference being that	except that
the question as to whether or not	whether
the reason is because	because
there are not many who	few
there is very little doubt that	doubtless, no doubt
through the use of	by, with
to be cognizant of	to know
to summarize the above	in summary
with a view to	to
within the realm of possibility	possible, possibly
with reference to	about
with regard to	concerning, about
with the (possible) exception of	except
with this in mind, it is clear that	therefore

Appendix III. Tautologies

Phrase	Improved Wording
absolutely essential	essential
adequate enough	either word, but not both
advance forward	advance
advance planning	planning
and etc.	etc.
attached together	attached
basic fundamentals	either word, but not both
circle around	circle

Appendix III.—Continued

Phrase	Improved Wording
circulate around	circulate
close scrutiny	scrutiny
collect together	collect
combine together	combine
completely full	full
consequent results	results
continue to remain	remain
desirable benefits	benefits
early beginnings	beginnings
enclosed herein	enclosed
end result	result
endorse on the back	endorse
entirely eliminate	eliminate
equal halves	halves
equally as good as	"equally" or "as good as", not both
fast in action	fast
fellow colleagues	colleagues
few in number	few
final completion	completion
first and foremost	first
first beginnings	beginnings
free of charge	free
important essentials	essentials
in the range of … to …	… to … (omit "in the range of")
join together	join
joint cooperation	cooperation
joint partnership	partnership
main essentials	essentials
merge together	merge
mix together	mix
mutual cooperation	cooperation
necessary requisite	one word, but not both
past experience	one word, but not both
plan ahead	plan
pooled together	pooled
rather interesting	interesting

Appendix III.—Continued

Phrase	Improved Wording
resultant effect	result
same identical	one word, but not both
surrounding circumstances	circumstances
ways and means	one word, but not both

Appendix IV. Absolute Words That Resist Modifiers

dead	mortal
extinct	obvious
fatal	peerless
final	perfect
honest	permanent
impossible	rare
inferior	safe
libelous	straight
lifeless	unique
matchless	universal
moral	vertical (or horizontal)

Chapter 3

Using Correct Punctuation

"Mother of Baby Strangled on Ship in Court"

This headline appeared in a local newspaper (*Rochester Times–Union*, August 5, 1986) and is definitely ambiguous, if not humorous. Was the mother strangled or was it the baby? Where did it happen? On ship? In court? Or was the ship in the court? Although newspapers need to minimize the number of words in headlines, this headline could have been made more clear with a change in word order and proper use of punctuation:

"Baby Strangled on Ship; Mother in Court"

This example of telegraphic writing used in newspaper and magazine headlines illustrates the importance of correct punctuation. Often, misplaced or missing punctuation can change the intended meaning of the writer's message. Let us look at some of the more common uses and abuses of punctuation that are found in technical writing. (For a more thorough coverage of punctuation rules, consult *The ACS Style Guide* [1] or any standard text on English grammar.)

Period

A period is one of the strongest punctuation marks. It indicates complete closure of a thought and signifies the end of an idea. Use a period at the end of a complete sentence.

1451–4/88/0021$06.00/1 © 1988 American Chemical Society

✔ Use a period with common abbreviations:

Mr.   Mrs.   Ms.   Dr.   Prof.

Many abbreviations used in scientific terminology do not require a period, especially those in the SI or metric systems:

g   mL   kg   mm   s

An extensive listing of acceptable abbreviations and symbols is included in *The ACS Style Guide* (1) and is not reproduced here.

Question Mark

✔ Use a question mark to end a direct question such as

Why was the burner turned off?

The question mark is a strong punctuation mark and denotes a complete ending of the thought or phrase. It is not used when posing an indirect question:

He asked why the burner was turned off.

✔ In a quotation, include the question mark inside the quotation marks if it is part of the quote:

He asked, "Why was the burner turned off?", and then relit it.

✔ If the complete sentence is a question, place the punctuation outside the quotation marks:

Why did she say, "It's cold in here", when it's 75 degrees?

Exclamation Point

An exclamation point normally has limited use in technical writing.

✔ Use an exclamation point to denote surprise or to signify that something unexpected, overwhelming, or unusual has occurred. It replaces a period and indicates a conclusion to a thought.

> It is claimed that Archimedes shouted "Eureka!" after he discovered a way to detect the amount of gold in the crown of the king of Syracuse.

> "Professor, I can't hand in my lab report because it completely dissolved in the nitric acid."
> "Well, that's the best excuse I've heard this week!"

Exclamation points, when used excessively, lose their effect and portray the author as a somewhat excitable or hysterical person.

Semicolon

✔ Use a semicolon to separate two independent clauses that are related closely. Independent clauses can also be written separately as complete sentences.

> The sodium chloride completely dissolved in the new solvent; the experiment was a success.

If the semicolon is being used correctly, the sentence could be rewritten by replacing the semicolon with a comma and the word "and".

> The sodium chloride completely dissolved in the new solvent, and the experiment was a success.

✔ Use a semicolon when adverbs (e.g., besides, furthermore, in fact, however, and otherwise) are being used as conjunctions.

> The class average was high; in fact, it represented the best performance since the class of 1984.

✔ Use the semicolon to separate items in a series or list if the items contain commas.

> The chemicals used in the experiment were potassium dichromate, $K_2Cr_2O_7$; ethanol, C_2H_5OH; and potassium hydroxide, KOH.

If there are no commas within the items you want to list, use commas to separate the items.

> The chemicals used in the experiment were potassium dichromate, ethanol, and potassium hydroxide.

Colon

✔ Use a colon to warn readers that they should pay attention to a list, explanation, or quotation that follows. A colon would be indicated in speech by a pause.

> He warned his colleagues: "Don't light a flame while I'm working with this chemical!"

Comma

The comma is probably the most abused punctuation mark from two aspects: some writers use it excessively, and some do not use it when it is necessary. A comma indicates a brief pause in the sentence and has many applications.

✔ In compound sentences, use a comma before conjunctions such as "and", "for", "nor", "or", and "but".

> The data points were collected in 2 days, but analysis was delayed for several weeks.

If the conjunction was omitted, the compound sentence could be broken into two complete sentences.

> The data points were collected in 2 days. Analysis was delayed for several weeks.

✔ When an incomplete phrase is used to introduce a sentence, set it off by a comma. The incomplete phrase cannot stand alone.

> Because benzene is carcinogenic, it is used less frequently as a solvent in the laboratory.

When the sentence is reversed, the comma is omitted.

> Benzene is used less frequently as a solvent in the laboratory because it is carcinogenic.

🖝 Use a comma to list a series of items when there is not internal punctuation in the list.

> I need to spend the evening writing my term paper for Chemical Literature, finishing my Physical Chemistry lab report, and washing the dishes.

Some sentences are ambiguous unless commas are inserted, in the same way that the headline at the beginning of this chapter was ambiguous.

> Except for the laboratory room space was limited.

This sentence could be interpreted in two different ways:

> Except for the laboratory, room space was limited.

> Except for the laboratory room, space was limited.

Apostrophe

🖝 Use an apostrophe to indicate a contraction or omission of letters. Do not confuse contractions with possessive pronouns: "it's" means "it is", but "its" is possessive; "who's" means "who is", but "whose" is possessive.

> It's going to snow tonight. (It is going to snow tonight.)

> The bird pulled its feather out.

🖝 Use an apostrophe to indicate ownership when a proper name is involved.

> Einstein's theory of relativity is what we will discuss today.

> Jones' car had two flat tires on the way to work today.

Quotation Marks

✔ Use quotation marks for direct quotations.

The professor said "Today is the last day to hand in your reports."

✔ If punctuation is not part of the material being quoted, place the quotation marks after the quote. (The ACS supports this style; other authorities always place the quotation marks after the punctuation. You should use the format that is followed by the journal, publisher, or organization for whom you are writing.)

I'm not sure who first called the phenomenon "acid-rain".

"Neat" is sometimes used to describe a liquid that is undiluted.

Direct quotations were not used in the preceding two examples, but quotation marks show that the words were being used in a special or new manner.

Summary

This overview of the correct uses of punctuation has been very brief. Any basic English grammar text will have a more thorough treatment of the subject and answers to questions about specific situations. In technical writing, the best rule to follow is to make the sentences as straightforward and simple as possible; in that situation, complicated punctuation rules can be frequently avoided.

Reference

1. *The ACS Style Guide*; Dodd, J. S., Ed.; American Chemical Society: Washington, DC, 1986; pp 42–69.

| Chapter 4 |

Selecting the Appropriate Verb

Verbs can be a major problem in technical writing, even though complex arrangements and usages are seldom necessary. Three common errors occur:

1. use of the wrong tense
2. use of the wrong voice
3. artificial creation of a verb from a noun

Tense

In scientific writing, a verb's tense (past, present, or future) has more importance than most people realize. The tense can convey implications that an author may not wish to make. For example, the statement "Einstein stated that mass was converted to energy" is grammatically correct, but use of the past tense "was" could be understood to mean that the statement is not true now. When the information is part of the established knowledge of the field, the present tense should be used. Previously published material is acknowledged in the same way. A better phrasing of the statement is "Einstein stated that mass is converted to energy" or "Einstein stated that mass can be converted to energy". The past tense is appropriate for the verb "state" because the work was done in the past; the present tense "is" tells the reader that the information is still valid.

1451–4/88/0027$06.00/1 © 1988 American Chemical Society

Usage

✔ In a typical scientific paper, use the present tense predominantly in the introduction where background information (previously established knowledge) is being given.

✔ Use the present tense for most of the content in the Discussion or Conclusion section.

✔ For the remainder of the paper, use primarily the past tense. The reason for past tense is that you are describing what you did (Experimental Section, Materials and Methods, or Measurements) and what you learned (Results) about the subject; it is not established knowledge because it has not been published or accepted by other members in the discipline. These guidelines will apply to most scientific papers, although occasionally a tense other than that specified will be appropriate.

Example

A good example of the changeover between tenses can be seen by taking isolated sentences from a recent paper, "Aqueous Coordination Chemistry of Vanadocene Dichloride, $V(\eta^5\text{-}C_5H_5)_2Cl_2$, with Nucleotides and Phosphoesters. Mechanistic Implications for a New Class of Antitumor Agents" (1) [Emphasis added].

The Abstract is a mixture of past and present tense. The past is used to indicate what was done: "An X-ray crystallographic study **was** also carried out", and the present tense is used for discussing and interpreting results: "The complex **crystallizes** in a monoclinic space group $P2_1$ (No. 4)."

The introduction cites existing knowledge in the present tense: "Kopf and Kopf-Maier have shown that metallocene dihalides and bis(pseudohalides) of the constitution $Cp_2MX_2(A)$. . . **are** highly active agents against Ehrlich ascites tumor (EAT) cells."

The Experimental Section (Methods and Materials or Physical Measurements) is presented predominantly in the past tense: "Organic solvents **were** thoroughly **dried** and **deoxygenated**. . .Infrared spectra **were recorded** on Perkin-Elmer 599 or 283 spectrometers."

The Discussion portion of the paper is primarily in the present: "The NMR relaxation studies **indicate** that near neutral pH, aqueous Cp_2VCl_2 **binds** selectively to the phosphate groups of nucleotides relative to the readily accessible nitrogenous sites on purine and pyrimidine bases."

The Conclusion also contains information given in the present tense: "The present results **indicate** that Cp_2MCl_2 interactions with DNA **are** likely to be very different in character than those of cisplatin." (The future could be used if suggesting further experimentation or recommendations.)

Voice

The selection of voice (active or passive) for a verb is an important factor to consider. In general, choose the active voice as frequently as possible. It is more direct, less ambiguous, less awkward, and easier to use than the passive voice. Many scientists resist using the active voice in writing. Part of this reluctance stems from the belief that scientific writing should be as impersonal as possible and that the best way to achieve impartiality is to use the passive voice. Currently, the consensus is that using "we" or "I" is acceptable and leads to a more straightforward presentation.

When making the choice between the active and passive voice, think about what you want to emphasize in the sentence. Typically, the subject receives the emphasis. Consider the following:

My car hit the tree.   (active)

The tree was hit by my car.   (passive)

Clearly, the first sentence is more direct and to the point. In the passive presentation, it sounds as if the tree were responsible for the accident! The active sentence is shorter than the passive by two words. Use of the active voice reduces the wordiness of a paper or talk.

Often the passive is the better voice to use when you want to emphasize your subject. For example,

The Warren Commission concluded that. . .Lee Harvey Oswald assassinated President Kennedy. (active)

or

> The Warren Commission concluded that. . .President Kennedy
> was assassinated by Lee Harvey Oswald. (passive)

An author who is writing about Oswald might prefer the active voice;
if Kennedy is the subject, the passive might be used to emphasize
him.

✔ Use the passive voice whenever the "acting source" or "agent"
is not known or is not important to the topic.

> The samples were maintained at 20 °C. (passive)

not

> The refrigerator maintained the samples at 20 °C. (active)

and

> The computer was stolen during the night. (passive)

not

> Someone stole the computer during the night. (active)

Business letters should not contain many passive constructions. They
often do, usually to conceal the active person or agent.

> It was decided that your position would be terminated. (Who
> decided?)

> Your $1,000,000 deposit was placed in the wrong account. (By
> what idiot?)

✔ Use the active voice when action is required or when specifying
instructions:

> The solution should be stirred until it turns green.

is not as effective as

> Stir the solution until it turns green.

and

> If my chemicals are not delivered tomorrow, your supervisor will be notified.

is weaker than

> If my chemicals are not delivered tomorrow, I will notify your supervisor.

Artificial or Contrived Verbs

The worlds of business and technology are uneasy about appearing ignorant. So it has been common practice to replace a simple verb with a complex one; when no complex verb exists, one is created. This practice of creating verbs can make sentences ambiguous, awkward, and often humorous. A student once wrote that "the sample was vacuumed for one hour" when the intention was to say that "the sample was kept under vacuum for one hour."

The computer areas are notorious for creating new verbs (not that other disciplines are far behind):

> He **keyboarded** the entries into the computer.

is a lot less clear than

> He **typed** the entries into the computer.

or

> He **typed** the entries on the keyboard.

Similarly,

> I **solutionized** the salt with water.

and

> This development will **impact** future efforts in this field.

are interesting, but

> I **dissolved** the salt in water.

and

> This development will **affect** future efforts in this field.

are better and use real verbs.

Reference

1. Toney, J. H.; Brock, C. P.; Marks, T. J. *J. Am. Chem. Soc.* **1986,** 108, 7263–7274.

Exercise

1. Select a journal article from your discipline and identify correct and incorrect uses of verbs: tense, voice, and artificiality.

ASSEMBLING YOUR REPORT

| Chapter 5 |

Planning a Paper or Talk: The Outline

O nly a few people have the ability to write a paper or give an oral presentation spontaneously. These lucky people probably have the ideas planned well in their minds. For most people, to have an effective paper or talk requires committing ideas to paper with considerable effort, and the hardest task is getting started.

The process of writing a paper is repetitious; it requires continued revising and editing. Figure 1 shows the six primary parts of the preparation process:

- focusing on the subject
- gathering information about the subject
- selecting the material to be used
- planning the sequencing and details of the paper
- writing
- revising

The lines connecting the steps in Figure 1 are meant to indicate that no particular order must be used for every attempt to produce a paper. After gathering information about the subject, you may wish to revise the way that you want to present the information or you may wish to alter the subject in minor or sometimes major ways. Similarly, you might start by preparing an outline for the paper, but after gathering information about the subject, you may want to amend it. Often

1451–4/88/0035$06.00/1 © 1988 American Chemical Society

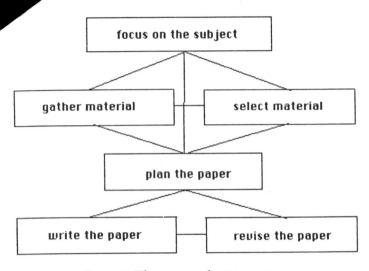

Figure 1. The process of writing a paper.

while putting the words onto paper, you may need to go back for more material and fill in gaps with specific information. The process is a fluid one requiring continued interchange of the various steps. Actually, all six steps should be connected by lines, but the most typical routes are the ones indicated in Figure 1.

The Planning Process

The best way to start is to simply write down your thoughts and intentions. Rewriting and revising the first draft, often several times, are normal (and desired) activities. Then you have a basis to consider, but it will seldom be in final form. Writing needs to be continually shaped and reshaped before it is presentable. This task of multiple revisions has been made considerably easier through the introduction of computer software and word processing devices. (*See* Chapter 10.)

The most tedious part of planning a paper, talk, or other presentation seems to be making an outline. Nevertheless, making an outline is probably the most important part of developing a good paper or talk. We have all had to listen to someone present a paper that he or she had not planned well. By taking the time to write an outline, the effectiveness of a paper or talk can increase by factors of 10 or more.

Most people who remember being forced to write outlines in elementary and high school regard this task as highly unpleasant. Perhaps this opinion results from never having been told how to create an outline as part of the complete process of preparing a presentation. The outline is not usually the first step, although it is often thought to be.

The first part of the planning process ought to include the answers to three questions:

1. What is the subject?
2. What is the purpose of the project?
3. Who will be the audience?

Subject

The response to the first question frequently seems obvious. A scientist writing a paper about the results of a research project knows exactly what subject to write about. A student is sometimes assigned a subject that should be addressed in a talk or paper. An employee might be asked (or required) to present information about a specific topic and would have no choice in determining the topic.

If you are given the subject, you still need to decide the focus. For example, suppose you pick the topic "gases" for a four- or five-page paper. Would your subject be abnormal gases, ideal gases, theories about gases, physical properties, chemical properties, or industrial uses of gases? After you pick one category and begin library research, you may still need to narrow your subject.

Purpose

Many times the purpose seems obvious. A scientist wishes to publish research results that are new or that add information to the body of knowledge in that science. A student is required to write a term paper to pass a course; sometimes the student even wants to write the paper because of interest in the subject! A manager might be planning an oral presentation to convince others about the worthiness of a project.

The preceding statements are actually rationales for beginning a project; they do not give the purpose of the undertaking. In considering the purpose, focus on what you will describe, what you want to prove, what you want to explain, or all three. Are you trying to provide information, trying to change someone's mind, or both?

Audience

Once you have defined your reasons for giving the talk or preparing the paper, you must consider your audience.

- Is your audience interested in the subject already or are you trying to make them interested?
- Do they already have an opinion about the matter?
- Are you trying to change that opinion or reinforce it?
- Will your audience consist of your superiors, peers, colleagues, or subordinates?
- Why should the audience want to get information from you and what qualifies you to provide it?
- What special information do you have that will be of interest to the audience?
- What is the background of the audience?
- If you are writing about a scientific or technical subject, are your readers likely to be lacking knowledge about the material or are they well aware of the topic?

Only when you have a clear idea of the answers to these questions will you be ready to actually begin the process of preparing the paper or talk. The type and level of language used, the amount of background information provided, and the tone of the language all depend on the purpose and the audience.

Writing the Outline

Now you are ready to begin the outline. Depending upon your familiarity with the subject, preliminary library research may be necessary before you can develop your focus on the topic. If, on the other hand, you are writing about research that you have worked on intensely on a day-by-day basis, you might feel qualified enough to begin the outline from information you have assimilated from your work.

Title

Start the outline by picking a tentative title. The title should be informative so that your audience knows precisely what subject you will discuss. Compose the final title after you write the paper because the focus of your topic may change as you gather and assemble information. For publications or presentations, the title is extremely important because the paper is often abstracted and indexed by key words that appear in the title.

For example, I recently presented a paper entitled "Hearing-Impaired/Deaf Students CAN Survive Chemistry!" (*1*) at a national American Chemical Society meeting. The title was indexed in the meeting abstracts as (key words are placed in boldface for emphasis):

ry. = **Hearing-**impaired/deaf students can survive chemist
Hearing- **impaired**/deaf students can survive chemistry. =
Hearing-impaired/deaf **students** can survive chemistry. =
students can **survive** chemistry. = Hearing-impaired/deaf

By trying to pick a flashy title, I may have lost potential listeners who looked for the key words "teaching", "handicapped students", or "deaf". In contrast, another paper given at the same meeting had a title that provided many key words (almost every noun and adjective):

"**Surface** Studies of **Dehydrofluorinated** PVF_2 **Films** Using **Photoacoustic** (PA) and **Attenuated** Total **Reflectance** (ATR) **FT–IR Spectroscopy**" (*2*)

Introduction

The next section of your outline should be about the introductory part of the paper or presentation. State the theme and purpose clearly so that the audience will know your intentions. These thoughts can be placed in a thesis statement or thesis sentence preceding the body of the outline, as shown in Figures 2 and 3. Occasionally you will need to state that portions of the topic will not be covered in the presentation and why those portions lie outside the scope of the paper.

Body of the Presentation

Your outline should then be concerned with the major thrust of the subject and the general and specific points that will be discussed.

Conclusion

Finally, your conclusion should summarize any arguments regarding various points in the body of the presentation. In addition, recommendations for action to be taken are often beneficial.

Format of Outlines

Preliminary

Outlines can have a variety of styles. When you are beginning to plan a presentation, the outline might only be a list of important topics that tentatively will be included in the presentation. For example, if you are planning to talk or write about "gases", you might start by listing the possible topics in a rough "scratch" outline. No particular format or order is necessary because a scratch outline is primarily for the person developing the material; no others are likely to see it. The following might be a rough beginning outline:

    I.   Properties of gases
    II.   Gas laws
    III.   Mixtures of gases
    IV.   History of discovery of gas laws
    V.   Theories about behavior of gases
    VI.   Real gases
    VII.   Deviations from ideal behavior

After doing some preliminary research, you would undoubtedly realize that the subject chosen is rather large for a typical paper or oral presentation. Then you would need to narrow the scope of the topic perhaps to only one of the items listed in the rough outline. You might realize that the arrangement of topics is not satisfactory. Outlines are made to be rewritten and should be revised continually as the paper takes form. The rough outline might be changed to:

    I.   Properties of gases
    II.   History of discovery of gas laws
    III.   Theory of behavior of gases
    IV.   Real gases and deviations from ideal behavior

When you have a good idea of the major topics to be covered in the paper, you will need to become more specific. By organizing the material well and in some detail, the actual writing process will become much easier.

Outline Styles

There are several ways to construct an outline, the most common of which is by topic. List information in brief phrases; do not worry about writing complete sentences. Figure 2 gives a typical example.

The Chemistry of Fluorine

Fluorine was difficult to isolate and identify because of its extreme reactivity, but this property also gives it important uses in chemistry and in industrial applications.

I. The element
    A. Isolation by H. Moissan in 1886
    B. Occurrence and natural abundance
       i. Minerals
          a. Fluorite
          b. Cryolite
          c. Fluorapatite
       ii. Sea water
    C. Physical properties, listed in a chart
    D. Chemical properties
       i. General reactivity trends
       ii. Representative bond energies
       iii. Interactions with solvents
II. Properties and preparation of compounds with fluorine
    A. Binary compounds
       i. With hydrogen
       ii. With metals
       iii. With nonmetals
    B. Ternary compounds
       i. Halogen oxide fluorides
          a. Oxofluorides [–OF]
          b. Peroxofluorides [–OOF]
       ii. Alkaline earth oxyfluorides
III. Production and uses of fluorine as an element
    A. Electrolytic production methods
    B. Production statistics in the United States
    C. Industrial uses
       i. Fluorinating agent
       ii. To make UF_6 for nuclear power plants
       iii. To make SF_6 for dielectrics
    D. Economic forecast for industrial production and usage

Figure 2. A topic outline.

It is difficult to construct an outline using complete sentences. You must have thought through the paper in detail and be solidly familiar with the subject. The benefit is that creating a sentence outline makes the actual writing of the paper or speech much easier. Figure 3 shows a possible sentence outline, adapted from the article "Food Irradiation: A technology at a turning point" (3) by Pamela S. Zurer.

For a very short paper, you might wish to make a paragraph outline rather than either of the preceding choices. The idea is to specify the contents of the information that will be presented in each paragraph of the final paper. For a one- or two-page paper, this can be done easily, but the task becomes formidable for longer efforts.

Labeling the Outline

In addition to the choice of using a topic, sentence, or paragraph outline, you need to decide whether to use the decimal system or the alphanumeric system when labeling the parts of the outline. The decimal system uses numbers for sections and subsections, as shown in Figure 3. Each new subsection retains the numbers of the statement above it and adds one more decimal number, starting with 1. For example,

    1. Major heading
        1.1 Subheading of topic 1
        1.2 Subheading of topic 1
            1.2.1 Subheading of topic 1.2
            1.2.2 Subheading of topic 1.2
                1.2.2.1 Subheading of topic 1.2.2, etc.

The alphanumeric system uses Roman and Arabic numerals and the upper and lower case English alphabet in alternating sequences, as shown in Figure 2. Another example of its use would be:

    I. Major heading
        A. Subheading of topic I
        B. Subheading of topic I
            1. Subheading of topic B
            2. Subheading of topic B
                a. Subheading of topic 2, etc.

A variation of the alphanumeric system uses upper- and lower-case

---

Food Irradiation: A Beneficial Use of Radioactivity

Food irradiation is a new technology that is used to preserve food by killing microorganisms and insects, but the use of radiation and its possible side effects on humans are controversial issues.

1. Food irradiation has had many beneficial applications.
   1.1 It inhibits the sprouting of potatoes and onions.
   1.2 It inactivates *trichinae* in pork.
   1.3 It delays ripening of some fruits.
   1.4 It kills microorganisms and insects in spices and seasonings.
2. The government and the general public have expressed concerns about the safety of food irradiation.
   2.1 Consumers are generally wary of the term "radiation".
      2.1.1 They associate it with cancer.
      2.1.2 The incidents involving radiation leakage at Three Mile Island and Chernobyl are fresh in their minds.
   2.2 The Food & Drug Administration (FDA) has approved the method at low doses (up to 100 kilorads).
      2.2.1 Because Congress defined radiation as a food additive in 1958, the FDA became involved.
      2.2.2 There are conflicting interpretations of the results of the FDA's testing.
         2.2.2.1 Tests were run on the radiolytic products.
         2.2.2.2 Some claim that the products are the same as those from cooking.
         2.2.2.3 Others think that each product should be tested.
3. Industries are reluctant to invest large sums of money into the technology until they are assured that the method is safe and that consumers will buy the irradiated products.
   3.1 The government has removed all regulatory barriers.
   3.2 Several industries have started to or plan to use the method.
      3.2.1 A New Jersey firm plans to irradiate pork.
      3.2.2 A Hawaiian firm plans to irradiate papayas.
   3.3 The FDA has already devised a "logo" to be placed on irradiated food.
4. Some of the same types of concerns were raised about other techniques (e.g., microwave) used for food and many of those methods are used routinely; there may be a future for food irradiation.

Figure 3. A sentence outline.

Roman numerals instead of Arabic numerals:

> I. Major heading
>   A. Subheading of topic I
>   B. Subheading of topic I
>   i. Subheading of topic B
>   ii. Subheading of topic B
>   a. Subheading of topic ii, etc.

Figures 2 and 3 show that certain conventions are followed for all methods. The tentative title of the paper is centered above the outline; it is not one of the headings, and it is not numbered or indexed. All new subheadings or subdivisions are indented, and if a sentence or a phrase is long enough to require a second line, it is indented.

The headings (and subheadings) should be specific and meaningful to the reader. A person should be able to get a good idea of the information that will be covered in a section simply by reading the headings. Below are two examples for the infamous typical essay entitled "How I Spent My Summer Vacation". The example on the left tells the reader almost nothing, whereas the outline on the right, although brief, has specific topics.

Poor Headings	*Better Headings*
I. How I spent my mornings	I. Mornings—yard work or house painting
II. How I spent my afternoons	II. Afternoons—surfing or reading
III. How I spent my evenings	III. Evenings—baby-sitting, parties, or television
IV. How I spent my weekends	IV. Weekends—outdoor camping
V. Conclusion	V. A profitable, enjoyable three months

It is also normal for headings using the same type of labeling (i.e., I, II, III; A, B, C; or i, ii, iii) to have roughly the same amount of importance or weight. This is true for subdivisions under one heading or subheading.

Consider the following examples of the major headings for a paper about the history of gas laws:

Unequal Weight	*Equal Weight*
I. Gas laws and discoverers	I. Gas law formulations
II. Boyle's Law	A. Boyle's Law
III. Ideal gases	B. Charles' Law
IV. Helium	C. Gay-Lussac's Law
	II. Ideal gas formulations
	A. Helium
	B. Other inert gases

In the left column, "Boyle's Law" and "Helium" are really examples of "Gas laws" and "Ideal gases", respectively, and should be subheadings of those major topics.

Parallel Form in Outlines

The style used in the headings (i.e., the words or phrases) should also be consistent, whether the outline is devised by using topics or sentences. This style is called parallel form and can be illustrated by looking at a portion of a paper about compounds of the halogens.

Nonparallel Form
I. Compounds with oxygen, hydrogen, and chlorine
  A. HClO
  B. Chlorous acid
  C. Hydrogen chlorate
  D. A very strong acid, $HClO_4$
II. $HBrO_x$ compounds

Parallel Form
I. Oxoacids of chlorine
  A. Hypochlorous acid
  B. Chlorous acid
  C. Chloric acid
  D. Perchloric acid
II. Oxoacids of bromine

A good outline has two or more entries for each subdivision or subheading. For every A, there must be a B; for every i, there must be an ii. If there is only a single entry, it generally should be incorporated into the heading for that category instead of being listed as a subtopic.

Poor Groupings
1.1 Origin of the universe
    1.1.1 Theories
1.2 Stellar evolution
    1.2.1 Spectral classes of stars
1.3 Synthesis of the elements
    1.3.1 Hydrogen burning
        1.3.1.1 Helium burning
        1.3.1.2 Carbon burning

There are usually several ways to solve the problem encountered in this example. An alternative arrangement retains logical groupings:

Logical Groupings
1.1 Origin of the universe
    1.1.1 Theories
    1.1.2 Stellar evolution
    1.1.3 Spectral classes
1.2 Synthesis of the elements
    1.2.1 Hydrogen burning
    1.2.2 Helium burning
    1.2.3 Carbon burning

or

1.1 Theories about the origin of the universe
1.2 Stellar evolution and spectral classes
1.3 Synthesis of the elements
    1.3.1 Hydrogen burning
    1.3.2 Helium burning
    1.3.3 Carbon burning

In these examples, no punctuation is used at the ends of the phrases. In a sentence outline, normal punctuation rules prevail. Capitalization is not used in a topical (or sentence) outline other than the first word of each heading (or sentence) and proper nouns.

Summary

Although these examples present the best way to outline, any outline is better than none. Attempting an outline will help you to organize your thoughts in a logical way and will enable you to present those thoughts better to others.

References

1. Cain, B. E. *Abstracts of Papers;* 193rd National Meeting of the American Chemical Society, Denver, CO; American Chemical Society: Washington, DC, 1987; CHED 53.
2. Salazar-Rojas, E. M.; Urban, M. W. *Abstracts of Papers;* 193rd National Meeting of the American Chemical Society, Denver, CO; American Chemical Society: Washington, DC, 1987; POLY 21.
3. Zurer, P. S. *Chem. Eng. News* **1986,** 64(18), 46–56.

Exercises

1. Prepare a topic outline about a scientific subject with which you are familiar (or about an assigned subject).

2. Read an article in a scientific journal and prepare a sentence outline of the article.

| Chapter 6 |

Gathering Data

Two major sources of raw data for a written report or oral presentation are (1) experimental results entered into a laboratory notebook and (2) previously published information obtained from library research. In both situations the material must be completely accurate and completely documented. Honesty and integrity are essential parts of a technical report, and all sources of data or information must be available for assessment and evaluation by colleagues.

The Laboratory Notebook

The laboratory notebook is one of the most useful tools for the professional scientist. It should contain a record of all work done in the laboratory, whether or not the work is considered successful. The notebook should be so detailed that another person could read it, follow the instructions, repeat the procedure, and obtain the same or similar results as the original experimenter. (Recent news headlines have documented cases where obtaining the same results has not been possible; the original experimenters wrote the desired results rather than those that they actually observed.)

Many beginning scientists believe that the laboratory book is simply a personal record, but it is more than that. Entries in manuals have played an important role in copyright and patent claims and have been used as evidence in legal matters involving product performance, discovery, and liability. For this reason, many industrial

1451–4/88/0049$06.00/1 © 1988 American Chemical Society

and academic research laboratories have developed policies of signing data pages or officially stamping them with the date and signatures to bear witness to the experimental data described.

Format

A professional record of laboratory work can be kept in various ways, and the ACS publication, *Writing the Laboratory Notebook* (1), is an excellent guide. Laboratory books for students should not be handled any differently than those for industrial and research scientists, although the format is flexible from one setting to another.

Some people prefer to arrange their records so that all rough data appear on one side of a page and the more refined treatment of the data (usually performed later) is on the other side. Another method is to use the back half of the book for rough, original data and the front half for the more formal presentation. Others prefer a strictly chronological recording of the data with refinements cross-referenced and inserted at the time they are actually done.

Basic Guidelines

Regardless of format, some basic elements should be followed:

✔ The name and address of the experimenter should be on the cover and on the inside of the book. Loss of data can be a huge problem if work must be repeated.

✔ The book should be bound.

✔ The first page(s) should be allocated to an index or table of contents and should be continually updated as entries are made.

✔ All pages should be numbered.

✔ All entries must be recorded in ink, not pencil.

✔ All entries must be dated.

✔ No entries should be obliterated, and no pages should be removed. Erasures and white-out paint are not permitted. Any entry that is thought to be wrong can be crossed out with a single line so that it can be read, and a notation should be made as to why it is considered erroneous. Too many times "dubious" information has actually proven to be valid and the information would have been lost if obliterated.

✔ All data should be entered directly into the book. Scraps of paper should not be used to record data.

✔ Everything should be noted, regardless of how insignificant or routine it may appear to the experimenter. Even opinions are encouraged in this circumstance (e.g., "The reaction appears to be progressing better at the temperature of 27 °C than it did at 37 °C.").

✔ Sketches, observations, rough graphs, and any other notations should be included.

✔ Other materials, such as computer output, spectra, chromatograms, and calculator tapes, should be incorporated into the laboratory book as mandated by the policy of the academic or industrial institution. Pasting and other means of entering this information are acceptable (1).

✔ Some professionals prefer a laboratory manual that has provisions for making a carbon copy of each page. The copy can be removed and stored in a secure place. This practice makes the possibility of losing vital data much lower because a copy is available.

✔ Computer systems are being used as alternate methods for keeping the records usually placed in the traditional handwritten laboratory book. As with any computer entry, a copy or duplicate (i.e., a backup diskette or disk) should be made of all entered information so that computer failure or user error does not result in the destruction or loss of the only copy.

Keeping good laboratory records is not difficult. Once the correct habits are learned, recordkeeping becomes a matter of daily routine; the word "daily" is important. Too many times, we decide to wait until tomorrow to write down a piece of information, and by tomorrow (or the next day), we have forgotten some of the important details.

The Library

Whether you are doing a report on original research or on a specified topic for a class or a job-related task, the most important resource for information is the nearest library. Searching the existing literature is the first step in gaining background on the subject. Searching the literature can help you narrow, refine, and define the scope of the research topic. If the task is to initiate a laboratory or experimental research project, it is absolutely necessary to find out what has already been done on the subject; it is a waste of time to "reinvent the wheel", and it is very valuable to learn what others have attempted or achieved in the area of interest. If you are writing a summary of existing knowledge, a search of the resources will assist you in focusing upon the correct aspect of the topic.

It sounds trivial, but you must become familiar with the layout of the library and the extent of its resources. I have never met a librarian who was not thrilled to show off the facilities to anyone who asked. If you are unfamiliar with the library, introduce yourself to the librarian and ask to be given a tour. If it is a large library, find out if one of the librarians specializes in the area in which you are interested. Chemistry librarians or science–technology librarians can be of tremendous assistance because of their detailed knowledge of the resources available in those disciplines.

Books

When beginning the search, try the card catalog first. Many libraries have the card catalog on microfiche, microfilm, or computers, but the information remains the same as the traditional entries on cards. Books, films, filmstrips, tapes, and other permanent resources are listed by subject, author, and title. Unless you are familiar with an author who has published books about your topic, the most logical place to begin the search is the subject index.

You might have to broaden or narrow your subject as you begin to look for references. For example, if you were considering doing a paper on AIDS, you would naturally first look for "AIDS" in the subject index. You would also check for information under "Acquired Immune Deficiency Syndrome", "Immunology", "Immune Resistant Diseases", "Diseases", "Epidemiology", and other related categories. Once you have found an available resource, locate it in the library (the call number will be on the card reference) and see if it is useful. Many times, although the book itself is not quite what you wanted, it may have useful references to other published material on the same subject.

The card catalog reference will contain all of the useful information about the publication, and you should record most of it for your files. The order will vary slightly depending on whether you are using the subject, title, or author index, but all will contain the same classifications:

- author's name (last name first) and often the date of birth (and death, if applicable)
- call number under which the book is to be found on the shelves

- title of the book
- edition (if more than one exists)
- city and state of publication
- name of the publisher
- year of publication
- number of chapters, number of pages, illustrations (if any), and height of book (i.e., X 250 p. illus. 21 cm)
- other technical information (e.g., Bibliography, pp 240–245)
- other headings under which the book is listed in the catalog, including the title
- Library of Congress call number (normally the same as the library's call number)
- Dewey Decimal System call number (sometimes used as the library's call number)

When searching the card catalog, do not neglect reference works that might be classified under categories such as "dictionary", "guide", "manual", "handbook", or "encyclopedia", because many of these works have been published in very specialized areas and they may be good starting points to lead you to other sources.

Periodicals

For most scientific and technological topics, you will find the most up-to-date information in periodicals rather than in books. For this kind of search, you will need to locate the various indexes for different types of periodicals. For popular publications, meaning those that the general public is likely to read, the *Reader's Guide to Periodical Literature* is the most useful. It is similar in organization to the card catalog and itemizes entries by subject, author, and title. Although the topic is scientific, the *Reader's Guide* should not be ignored; many publications have portions devoted to science and technology, and often these articles are understood more easily than those found in the strictly professional journals. (The reason is that a technical writer has adapted the information for public consumption!)

Professional journal articles and other publications can be easily located through the use of abstracting and indexing services. The abstracting volumes provide a synopsis of the article, which helps the reader to decide whether the original article should be pursued. Indexing services generally do not provide more than the author, title, and journal in which the article was published. Nevertheless, both are quick ways to find out if recent scientific information has been published related to your topic.

Because volumes of indexes and abstracts are relatively expensive and many libraries are suffering from restricted budgets, libraries tend to maintain only those volumes that are particularly useful to their clientele. You may need to visit several libraries in a large city or go to specialized libraries at universities or colleges. The reference librarian can tell you where the nearest required resource is housed.

Indexes and Abstracts

The following are brief descriptions of the most common volumes of use to science–technology fields. Some indexes are devoted to very specific topics and are not included in this list; again, ask the reference librarian or consult one of the resources listed at the end of the chapter.

- *Applied Science and Technology Index*: A pointer to publications that may be of interest in industrial and technological applications of science.

- *Biological Abstracts*: Comprehensive abstracting of articles in the biological and life sciences.

- *Chemical Abstracts*: The most comprehensive coverage of chemical publications. Indexes are arranged by subject, author, title, compound name, formula, and registry number. Journal articles, patents, reviews, papers, dissertations, and other reports are included.

- *Chemical Titles*: A reproduction of the tables of contents from about 700 journals related to chemistry. Indexes by author and key words from the titles are provided.

- *Current Chemical Reactions*: Lists new synthetic methods by name, type of reaction, starting materials, products, reagents, and catalysts. Also has indexing by corporation, author, and journal.

- *Current Contents*: Includes key words from titles and has different sections dealing with physical, chemical, and earth sciences; life sciences; and engineering, technology, and applied sciences.

- *Engineering Index*: Brief abstracts on topics of engineering with a broad coverage of subjects.

- *Index Chemicus*: An index to new organic compounds and their chemistry. It has indexing by molecular formula, name, biological activity, author, labeled compounds, journal, and corporation or organization.

- *Index Medicus*: Lists articles from medical and biomedical journals.

- *Physics Abstracts*: Coverage of publications in research areas related to physics.

- *Science Citation Index*: Enables the user to look for a specific author and learn who has cited that author in a publication. This index gives you one way to identify authors who are working in the same subject area.

A wide variety of other publications may provide information that overlaps with your topic:

Agricultural Index	*Environmental Index*
Computer and Control Abstracts	*General Science Index*
Education Index	*Pollution Abstracts*
Electrical Engineering Index	*Psychological Abstracts*
The Energy Index	*Sociological Abstracts*

Computer Searches of the Literature

Disciplines such as chemistry have had an exponential increase in the number of primary publications in professional journals. With this dramatic increase, it is nearly impossible for anyone to keep abreast of the total literature of a discipline, and it requires effort and time to stay current even in a very narrow field. Add to this the fact that many libraries have not been able to maintain subscriptions to all of the indexes and abstracts that they might have supported in the past, and we see the need for the use of computers in literature retrieval.

Fortunately, many of the abstracting services have computerized their publications, and these have been entered as data bases for on-line access by subscribing customers, such as libraries and industries. Access to millions of citations in very specialized areas is made possible through interconnecting data bases. Results of a search can often be printed on-line immediately, although most systems have a provision for off-line printing, which is considerably cheaper if the user is not in a tremendous rush.

Searches can be costly because of the immense amount of data available, and you must narrow the subject sufficiently to eliminate articles that would not be useful. If you want to get the most from a search, reference librarians request that you be prepared by doing the following:

- Do some preliminary searching among the abstracts and indexes available in your library. Get an idea of how much information is available and how you can limit your search to the immediate area of concern.

- Read or scan several of the references that you have found. Maybe these will help you to clarify your needs.

- Write a short (25–50 words) summary or description of your topic. This summary will assist the reference librarian and it will also force you to define your subject. If it is difficult to write this description, you are prematurely considering yourself to be ready for the computer search.

- Underline the key (important) words in your description and decide which are most important. Be careful on your choice; most search systems use a type of Boolean logic.

If you wanted to write about the chemistry of blue jeans (denim) in connection with indigo dye and the plant from which it is obtained, *Indigofera tinctoria*, you have several choices in narrowing the search. Choosing blue jeans OR indigo dye OR *Indigofera tinctoria* would result in a listing of every paper about blue jeans (whether it mentioned

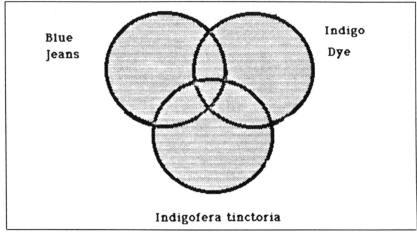

Figure 1. Jeans OR dye OR plant.

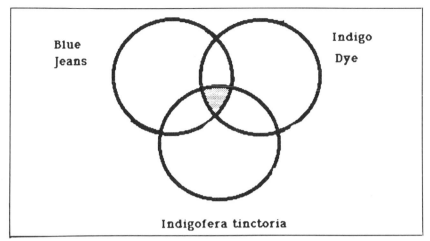

Figure 2. Jeans AND dye AND plant.

the dye or the plant or not) and every paper about indigo dye (whether it mentioned blue jeans or the plant) and every paper about *Indigofera tinctoria* (whether it mentioned the blue jeans or the dye). Such listings would provide too much information and would not be very helpful. The shaded section in Figure 1 represents the areas searched.

Selecting blue jeans AND indigo dye AND *Indigofera tinctoria* might be too restrictive because it would list only those articles that contained information on all three items, as shown in the shaded section in Figure 2.

Perhaps you have decided that the most important part is the indigo dye, and the other two topics are secondary. You might want to select indigo dye AND (blue jeans OR *Indigofera tinctoria*); in that case, you would receive a listing of all papers that mentioned indigo dye and blue jeans or indigo dye and the plant, as shown in the shaded part of Figure 3.

If the original search does not provide enough resources, try to think of appropriate synonyms to use. For example, blue jeans might also be listed separately in some entries as "denims".

Using the Sources and Taking Notes

Once you have found appropriate books and journal articles, it is time to use them. The presence of photocopiers in most libraries has made it easy and convenient to make permanent records of information

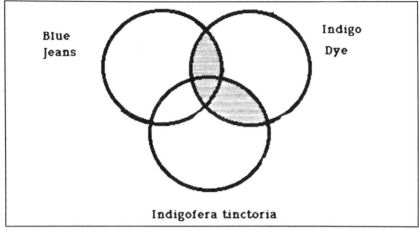

Figure 3. Dye AND (jeans OR plant).

pertinent to a research topic. If photocopying is the method used, it is important to write all of the bibliographical data on the photocopy so that it will be easy to obtain if you wish to cite or quote the work in your final report. In many scientific journals, most of the necessary information appears at the top or bottom of every page, but make sure that all you need is present; it is tedious to have to look up the reference a second time when you are missing an item for a footnote or bibliography.

In addition to the prohibitive expense, it is a violation of copyright laws to photocopy large portions (or all) of a book. Individual pages would be acceptable, but it is preferable to invoke the old-fashioned art of making notes. It is best to make notes in an organized way to make the task of combining information later a little easier. People have their own techniques for jotting down notes, and the best system to use is whatever works for you.

Use of index cards is prevalent among those who do a lot of library research. In contrast to entering information into a bound notebook, index cards offer the flexibility of being rearranged in various sequences as the text of the paper goes through the revisions that always occur.

Before making notes, read the article or book chapter quickly for an overview of the information contained. Summarize the material in your own words (paraphrase) and write it down with all of the pertinent information about the source in which you found it. If you find statistics, catchy expressions, or other information that you think you might want to use, write it down exactly as it appears in the source and put quotation marks around it so that you know later that it was not your paraphrase. Be sure to indicate the exact page on which the material was found. Some people make one card for the source and then put the helpful bits of information on other cards, using the author's name as a cross-reference.

Any method that will provide you with the information you will need later is appropriate. You must, however, be sure to give credit to the source from which you obtained the data (see Chapter 7).

Reference

1. Kanare, H. M. *Writing the Laboratory Notebook*; American Chemical Society: Washington, DC, 1985.

Bibliography

Abstracting and Indexing Services Directory; Schmittroth, J., Jr., Ed.; Gale Research: Detroit, 1982–1983.

Malzell, R. E. *How to Find Chemical Information*; Wiley: New York, 1987.

Owen, D. B.; Hanchey, M. M. *Indexes and Abstracts in Science and Technology: A Descriptive Guide*; Scarecrow: Metuchen, NJ, 1974.

Exercises

1. Select a topic of interest (or one assigned) and list the key words you would use to locate articles of interest.
2. After completing 1, use the abstracts in your discipline to identify journal articles and other publications related to your topic that were published during the past 5 or 10 years. Do you need to select new key words? Why or why not? Keep a list of the articles found for preparing a bibliography (see Chapter 7).

| Chapter 7 |

Documenting: Footnotes, Endnotes, and Bibliographies

Every scientific article requires documentation of where (or from whom) the source material was obtained. This documentation is a matter of courtesy to the author or researcher who is the source of information, and it is a matter of professional honesty to give credit where credit is due. Formal laboratory reports may require references to various techniques or protocols.

Three major ways of providing this information are in footnotes, endnotes, or bibliographies. In many applications, combinations of these may be used. Several formats in which this information is given and examples will be shown later in this chapter. Regardless of the format used, the most important fact to remember is that any person reading the material must be given complete references so that he or she could go to the original source.

If you are writing an article for submission to a scientific journal, look at published articles in the specific journal being considered and use the styles shown as models for your paper. Most journals publish a "Guidelines for Authors" or "Information for Authors" section in the first (or last) volume of the year. Find the most recent copy of the guidelines for that journal and follow the instructions.

When you are gathering material for a journal article, make sure that you have complete documentation. You will need to record the author(s)'s name(s); title of publication; name of journal; publisher and site of publication (book); volume number; year of publication;

1451–4/88/0061$06.00/1 © 1988 American Chemical Society

and pages, chapters, or sections as appropriate. All of the preceding facts might not be required, but if for some reason the journal selected does not accept the proposed paper, you may wish to submit it to another journal that could require a different set of information in its format for documentation. Nothing is more bothersome than being forced to find all of the reference materials again to obtain the additional information!

In the past, footnotes (or endnotes) and a bibliography were both required for a formal paper. This requirement resulted in considerable redundancy because any source referenced with a footnote also had to be listed in the bibliography. However, a bibliography could, and often did, contain other sources of information that were not used directly. Because the costs of printing have escalated, many journals avoid this duplication by using a variety of acceptable methods that will be discussed.

Before discussing the specific styles, let us consider first what a footnote is and when to use one. Two types of footnotes are common: expository and references.

Expository Footnotes

An *expository footnote* generally contains material that supplements information contained in the body of the text. Normally, you should try to assimilate this information into the article itself, but sometimes doing so destroys the flow of the text. The following might be an excerpt from an article that uses expository footnotes:

> . . . and sold 5.5 million tons[1] of grain to the USSR before it was realized that DDT[2] caused environmental and health problems. Samples of the grain were analyzed[3] and found to contain. . .

---

[1] A ton is defined as 2,000 pounds.

[2] For information about the discovery of DDT, see Reference 5.

[3] All sample analyses were performed by the AB Chemical Consulting Service, Buffalo, NY.

As indicated in the example, expository footnotes can be used to clarify, to provide additional information, and to make a peripheral comment. They can be used to cross-reference other portions of the text or report. An expository footnote is always placed at the bottom of the page on which it is referred so that a reader can quickly scan to see if the information is pertinent to understanding the material.

When used to provide a reference to a literature source, the information may be placed at the bottom of the page (a footnote) or at the end of the article (an endnote).

References

References to the material in the text are required for all direct quotations, summaries or paraphrases of someone else's opinions or ideas, precise statistical information, and precise factual information.

If a direct quotation is used in the text, it must be documented. Often people have difficulty deciding when they must acknowledge a phrase as a quotation. The general rule is that if you have used four or more consecutive words of someone else's writing, the words must be quoted. For example, if someone was writing an article on the subject of plagiarism, using this book as a source of information, he or she would have to use a quote in the following paragraph:

> *Plagiarism is one of the most unnecessary academic errors. In every case, it can be avoided and, in many cases, it occurs because of the author's carelessness or ignorance. A recent publication states that the rule is simply that when you use "four or more consecutive words of someone else's writing, the words must be quoted (4)."*

Obviously, common sense must be used when following this guideline. We do not acknowledge Washington or Jefferson every time we use "United States of America", and we do not need to reference other phrases that are part of everyday use. If the matter in question is common knowledge or part of your normal vocabulary, it will not be referenced or placed in quotation marks in the sources you are using. Another way to determine whether it is general knowledge is to ask yourself if the information is something that you could casually introduce into the luncheon conversation with friends; if it is not, you should document the information.

Another general rule is that if the quote contains more than 200 words, you need permission from the publisher to reproduce it.

In addition to direct quotations, paraphrasing of information from another source should be documented, especially when using someone's data or assertions to support your thesis. For example, stating that "95% of red-haired children have green eyes" certainly needs to be referenced. A reader may not believe that statement and would wonder where the author obtained the (erroneous) information. Statements that are generally accepted, such as "DDT causes a detrimental effect on the hatching of birds' eggs", have much more strength if documented.

A scientific paper normally will have an abundance of footnotes (or endnotes). A paragraph based on others' information should be referenced to indicate the source of the ideas. (On occasion, people use references to establish their credentials or for prestige. Citing a famous scientist or a classic paper is one method used to show that the author of the paper is putting herself or himself in the same stratum.)

Referencing Styles

As noted previously, journals and publishers vary in the formats that they require for their publications, and an author should conform to the specifications provided. Three common ways of indicating a reference are

1. The crystal structure of ammonium sulfamate has been determined[5].

2. The crystal structure of ammonium sulfamate has been determined (5).

3. The crystal structure of ammonium sulfamate has been determined (Cain and Kanda, 1972).

---

5. Cain, B. E.; Kanda, F. A. *Z. Kristallogr.* **1972,** *135,* 253–261.

If several sources are referenced in one location, commas (or semicolons) are used to separate the numbers, or dashes can be used if three or more consecutive sources are cited:

The crystal structure of ammonium sulfamate[5,6]. . .

The crystal structure of ammonium sulfamate (4–6). . .

The crystal structure of ammonium sulfamate (Cain and Kanda, 1972; Jones, 1981).

A citation does not have to be at the end of a phrase or sentence; quite often it is appropriate at the actual point of mention, especially if a sentence contains multiple citations. The author(s)'s name(s) could appear in the text as a variation in technique, and this variation is often done if the name, title, or reputation of the person strengthens the material presented:

According to Cain and Kanda[5], the structure of. . .

The structure was determined by Cain and Kanda[5] and later refined by Jones[6] . . .

Nobel Prize recipient Linus Pauling[7] believes that vitamin C. . .

Footnotes should appear at the bottom of the first page on which the number is used, and they should be listed in numerical order. Endnotes should also be in numerical order, but on a separate sheet at the end of the paper. If the referencing uses names of the authors, then the sources should be listed at the end of the paper in alphabetical order by the first author's surname.

Journals

For journals, the footnote should contain the names of the authors, the name of the journal, volume number, year of publication, and page number. Normally, initials are used for the author's first and second names. The title of the journal is typically abbreviated unless it is one word; standard accepted abbreviations can be found in the *Chemical Abstracts Service Source Index (CASSI)*. The journal title and volume number are placed in italics; in a typed manuscript, they would be underlined. The year of journal publication is in boldface, which is indicated in the typed paper with a wavy underscore. For example,

Typeset

8. Ashby, M. T.; Enemark, J. H.; Lichtenberger, D. L.; Ortega, R. B. *Inorg. Chem.* **1986,** *25,* 3514.

Typewritten

> 8. Ashby, M. T.; Enemark, J. H.; Lichtenberger, D. L.; Ortega,
>   R. B. <u>Inorg. Chem.</u> <u>1986</u>, <u>25</u>, 3514.

Many journals request that the page numbering be inclusive; in the previous example, the last entry would be 3514–3517. For a term paper or in-house report, it may be desirable to include the title of the article in quotes between the author(s) and the journal name:

> 8. Ashby, M. T.; Enemark, J. H.; Lichtenberger, D. L.; Ortega,
>   R. B. "Acyclic Polythioether Complexes: Preparation and
>   Crystal Structure of Tricarbonyl-(2,5,8-trithianonane) molyb-
>   denum(0)". *Inorg. Chem.* **1986,** *25,* 3514–3517.

If the journal does not have sequential page numbering from one issue to another within a volume, indicate the issue number by placing it in parentheses after the volume number. The date of the issue could be used, although the issue number is preferred.

> 9. Baum, R. *Chem. Eng. News* **1987,** *65(12),* 21–22.

or

> 9. Baum, R. *Chem. Eng. News* **1987,** *65,* March 23, 21–22.

The older style of referencing journals is still being used; this style places the year of publication in regular type at the end of the footnote in parentheses and separates the author(s)'s name(s) by commas rather than semicolons. The volume number would then appear in boldface immediately after the journal name.

> 8. Ashby, M. T., Enemark, J. H., Lichtenberger, D. L., Ortega,
>   R. B. *Inorg. Chem.,* **25,** 3514(1986).

Books

The information given when citing a book is similar to that for a journal: name of author(s), title (in italics), publisher, city and state, date of publication, and page (or chapter). For example,

Typeset

> 10. Perrin, D. D.; Armarego, W. L. F.; Perrin, D. R. *Purification of Laboratory Chemicals*; Pergamon: New York, 1980; p 5.

Typewritten

> 10. Perrin, D. D.; Armarego, W. L. F.; Perrin, D. R. <u>Purification of Laboratory Chemicals</u>; Pergamon: New York, 1980; p 5.

Variations include the following:

> 10. Perrin, D. D., Armarego, W. L. F., Perrin, D. R., "Purification of Laboratory Chemicals", Pergamon, NY (1980), p 5.

> 10. Perrin, D. D., Armarego, W. L. F., Perrin, D. R., "Purification of Laboratory Chemicals" (NY: Pergamon, 1980), p 5.

If the book has an editor, the format would be

> 10. *The ACS Style Guide*; Dodd, J. S., Ed.; American Chemical Society: Washington, DC, 1986; pp 110–112.

If a chapter is written by someone other than the editor,

> 12. Antony, A. A. In *The ACS Style Guide*; Dodd, J. S., Ed.; American Chemical Society: Washington, DC, 1986; Chapter 6.

Abbreviations Used in Footnotes

In addition to the abbreviated name of the journal, several other standard abbreviations can be encountered in footnotes, endnotes, or bibliographies, including some that are no longer in style. In citing the publisher of a book, normally the words "Company", "Inc.", "Publisher", and "Press" are not used. You will notice in the previous examples that the abbreviations for page (p) or pages (pp) are not followed by periods. For an editor (or editors), "Ed." (or "Eds.") would be used, but "ed." is proper for edition. When referring to sections of books, the words "Chapter" and "Part" are used completely whereas "Volume" is abbreviated as "Vol."

Latin words were used extensively in footnotes in the past, but they are now considered by many to be pedantic and are frequently replaced by their English equivalents. Nevertheless, you should be aware of their meaning and usage because you will continue to see them in published material. The most common abbreviations used in the text are set in regular Roman type, although some publications continue to italicize all Latin or non-English words.

One of the most common is "Ibid." (ibidem, "in the same place"). "Ibid" is used if the reference immediately following a cited work is to the same source.

> 10. Perrin, D. D.; Armarego, W. L. F.; Perrin, D. R. *Purification of Laboratory Chemicals*; Pergamon: New York, 1980; p 5.

> 11. Ibid., p 10.

Instead of using "ibid." for a previously cited source, it is now common to list only the author's last name and the page number. If there are many different entries by the same author(s), they should be differentiated by inserting the title or year, as appropriate.

> 10. Perrin, D. D.; Armarego, W. L. F.; Perrin, D. R. *Purification of Laboratory Chemicals*; Pergamon: New York, 1980; p 5.

> 11. Perrin et al. p 10.

You will notice the use of "et al." (et alii or et aliae, "and others") for multiple authors.

"Op. cit." (opere citato, "in the work cited") can be used to indicate a previously cited work that does not immediately precede the footnote; it should not be used if the original citation is distant from the present one.

> 10. Perrin, D. D.; Armarego, W. L. F.; Perrin, D. R. *Purification of Laboratory Chemicals*; Pergamon: New York, 1980; p 5.

> 11. *The ACS Style Guide*; Dodd, J. S., Ed.; American Chemical Society: Washington, DC, 1986; p 110.

12. Perrin et al. Op. cit., p 6.

If only one work is cited in the paper, "op. cit." can be replaced simply by using the author(s)'s name(s) and page number, as in the example given after the discussion of "ibid." If you are citing multiple articles by the same author(s), the articles must be identified by including the title or year.

Occasionally you will also see "q.v." (quod vide, "which see") used in an expository reference.

13. There is a very good account of this subject, q.v. Ref. 6.

"Q.v." can be replaced by its English equivalent:

13. For a very good account of this subject, see Ref. 6.

Other familiar Latin abbreviations are generally used in the body of the text rather than in footnotes. They are normally not italicized when used: "i.e." (id est, "that is to say") and "e.g." (exempli gratia, "for example"). Both are followed by a comma (e.g., as in this example). The data section of a paper might also contain "ca." or "c." (circa, "about") in a phrase such as "we used a small amount (ca. 0.2 mL)."

Bibliographies

A Bibliography (or References or References Cited) is normally the last page of a paper or book chapter, and the entries have the same format as footnotes, with the exception that the titles of journal articles are usually included. The list is arranged alphabetically by the first author's surname or by book or journal title (ignoring "a", "an", and "the") if no author is indicated.

The bibliography must contain all sources that have been used in footnotes or endnotes. Additionally, it usually contains other sources that were consulted during the library research but were not cited directly in the paper or book. If only certain chapters, pages, or volumes of the source were used, they should be indicated. If the method of referencing used in the paper was by author name and date, the bibliography can serve as the endnotes unless otherwise specified. A short bibliography for a paper about the Chernobyl nuclear accident is shown in the box.

Bibliography for a Paper About the Chernobyl Nuclear Accident

1. Allman, W. F. "Chernobyl: An Overreaction?"; *Science 86* **1986,** 7(6), 11.

2. Greenwald, J. "And Now, the Political Fallout"; *TIME* **1986,** June 2, 47–49.

3. Greenwald, J. "Deadly Meltdown"; *TIME* **1986,** May 12, 39–52.

4. Greenwald, J. "More Fallout from Chernobyl"; *TIME* **1986,** May 19, 44–46.

5. Marshall, E. "The Lessons of Chernobyl"; *Science* **1986,** 233, 1375–1376.

6. Norman, C. "Chernobyl: Errors and Design Flaws"; *Science* **1986,** 233, 1029–1031.

7. Norman, C. "Hazy Picture of Chernobyl Emerging"; *Science* **1986,** 233, 1331–1333.

8. Norman, C.; Dickson, D. "The Aftermath of Chernobyl"; *Science* **1986,** 233, 1141–1143.

9. Powers, T. "Chernobyl as a Paradigm of a Faustian Bargain"; *Discover* **1986,** 7(6), 33–35.

10. Serrill, M. S. "Anatomy of a Catastrophe"; *TIME* **1986,** 128(9), 26–29.

11. Serrill, M. S. " 'We Are Still Not Satisfied' "; *TIME* **1986,** 128(10), 46.

12. Sherman, A.; Sherman, S. J. "Nuclear Power", *Chemistry and Our Changing World*; Prentice–Hall: Englewood Cliffs, NJ, 1983; Chapter 11.

13. "Soviet Nuclear Accident: U.S. scrutinizes fallout, other effects"; *Chem. Eng. News* **1986,** 64(19), 4–5.

14. "There is a Silent Enemy Lurking"; *TIME* **1986,** June 23, 49.

Exercises

1. Using the information and format from a journal in your scientific area, document this book and a journal article as a footnote.

2. Search through a book or journal and find an expository footnote. Copy the paragraph that contains the footnote and document the source of the paragraph.

3. Determine what is wrong with the following footnotes and correct them to correspond to the preferred style given in the chapter:

   1. Alan Sherman and Sharon J. Sherman, *Chemistry and Our Changing World*, Prentice-Hall, Inc., Englewood Cliffs, NJ; 1983; p. 22.

   2. Sime, R. L. *J. Chem. Ed.*; **63,** 1986, 653–657.

   3. Kerr, R. A. *Sci.* **1984,** 226, pp. 954–955.

| Chapter 8 |

Supplementing with Visual Aids and Graphic Art

Frequently in technical presentations, written or oral, material is most effective in a visual format. A clear diagram, chart, or graph can be a positive supplement to the text or speech and, occasionally, the visual aid can be a stand-alone piece of information.

Whether the visual aid is to be used as a slide or transparency or as part of a printed report, document, or journal article, several guidelines should be followed to enhance its use:

✔ The visual aid must be clear to the reader or the listening audience. You have indeed been fortunate if you have managed to avoid a presentation where the speaker presents a slide that is packed with information to the extent that it looks like a page from a telephone directory. The speaker then inevitably says, "As can easily be seen from the data. . ."! When preparing or planning slides or transparencies, you must consider the size of the room, location of projection equipment, and anticipated size of the audience itself. If the visual aid is to be printed, you need to know how much the material will be reduced in size (if at all), and you must design the lettering, numbers, and diagram so that everything will be legible at that size. The print and numbers should not vary much in size because reduction may distort one more than the other.

✔ If the visual aid must be cluttered or "busy", highlight the important information by enlarging those specific words or numbers or by using color, underlining, or arrows. With the increasing availability

1451–4/88/0073$06.00/1 © 1988 American Chemical Society

of photocopiers that can duplicate colors, these effects are becoming easier to achieve. However, if you use colors to excess, you lose the desired effect. Do not use too many colors, because the viewer may not be able to decide which information is most important. In addition, if different colors are used, they must look different; use of red and orange together is not as distinctive as red and blue would be. Contrast is important in considering the use of a colored background. Yellow letters with blue backing may be less legible than the standard black printing on white.

Types of Visual Aids

Types of visual aids can be varied as appropriate: tables, graphs (bar and line), charts and figures, diagrams, photographs, spectra, flow charts, and sketches all have their uses.

Tables

A table is one of the easiest forms of presenting information, and probably, because of its ease, is one of the most abused. For example, Table I provides chemistry exam scores that might have importance to the instructor but are not particularly easy for the casual viewer or reader to interpret. If the report would be incomplete without the individual listing of scores, make the list an Appendix for referral purposes only.

A better presentation of the same information can be seen in Table II. In this example, only twelve grade ranges need to be considered, and percentages have been included to make the interpretation more clear. Traditional ranges incorporating a ten-point spread have been used, and apparently the instructor wished to emphasize the grades of 0 and 100 by isolating them in the table.

Because the data given are exam scores, Table III might be a more appropriate way to display them; that is, by letter grade assignment. Table III is more concise than Table II and certainly emphasizes that the class performance on that particular exam was skewed to the low end of the grading system.

Although the examples are tables of numerical data, a tabular arrangement of material from the text can also be used. Again, the deciding factor should be whether the table assists the reader or viewer

Table I. Scores on Exam 1 in Chemical Foundations

Score	Number of Students	Score	Number of Students
0	10	51	2
1	0	52	3
2	0	53	1
3	0	54	2
4	0	55	2
5	0	56	5
6	0	57	3
7	0	58	8
8	0	59	4
9	0	60	4
10	0	61	2
11	0	62	4
12	0	63	2
13	0	64	3
14	0	65	2
15	1	66	1
16	0	67	2
17	1	68	6
18	0	69	3
19	1	70	3
20	1	71	4
21	0	72	4
22	0	73	0
23	2	74	3
24	1	75	3
25	0	76	2
26	0	77	3
27	0	78	1
28	1	79	5
29	1	80	4
30	0	81	2
31	0	82	0
32	1	83	3
33	2	84	9
34	1	85	3
35	0	86	3
36	4	87	3
37	1	88	3
38	1	89	4
39	1	90	3
40	0	91	2
41	0	92	2
42	1	93	2

Table I.—Continued

Score	Number of Students	Score	Number of Students
43	2	94	1
44	1	95	0
45	2	96	1
46	3	97	5
47	0	98	0
48	2	99	2
49	4	100	7
50	2		

Table II. Range of Scores on Exam 1 in Chemical Foundations

Range	Number of Students	Percentage of Total ($N = 192$)
0	11	5.7
1–9	0	0.0
10–19	3	1.6
20–29	6	3.1
30–39	11	5.7
40–49	15	7.8
50–59	30	15.6
60–69	29	15.1
70–79	28	14.6
80–89	34	17.7
90–99	18	9.4
100	7	3.6

Table III. Grade Distribution for Exam 1 in Chemical Foundations

Range	Number of Students	Percentage of Total ($N = 192$)	Letter Grade
90–100	25	13.0	A
80–89	33	17.2	B
65–79	41	21.4	C
50–64	47	24.5	D
0–49	46	24.0	F

to interpret or assess the information or whether there is a more effective way of portraying the same data or text.

Graphs

Graphs should be considered as viable means to present numerical results. Many times a trend in the data can be detected much more effectively in a graph. Two common types of graphs from which to choose are the bar graph and the line graph.

A bar graph is especially useful in looking at selected blocks of data and would be a good alternative to use with the grades from Tables I–III. Because a bar graph often involves percentages, the scale used is critical. The information in Table III shows that the highest percentage involved in plotting the grades would be the D student group of 24.5%. If a range of 0–100% was selected when constructing the graph, about three-quarters of the paper would be empty. By using a range of 0–30% (maximum), the graph could be expanded to fill the entire page. Figures 1 and 2 compare the use of the two ranges.

A bar graph can be used effectively to compare or contrast groups of results from multiple experiments, years, or events. A key must be provided to differentiate the two sets of data, as shown in Figure 3. Normally, the bars used should be the same width, and the differentiation should be indicated by pattern or color.

Line graphs are also useful in showing trends or changes. Some basic rules that must be followed when preparing them are the following:

✔ The axes must be labeled with a description of the measurement and the appropriate units, if any. For example, a Beer's law plot (spectrophotometric determination) might have the y-axis labeled as "Absorbance, A" and the x-axis labeled as "Concentration of $KMnO_4$, M".

✔ The graph must have a title or caption that should not repeat what is plotted on the axes; "Absorbance versus Molarity of $KMnO_4$" would not be an acceptable title because anyone would see that if the axes are correctly identified on the graph. A better title would be "Beer's law plot for $KMnO_4$" or "Visible spectrophotometric determination of $KMnO_4$".

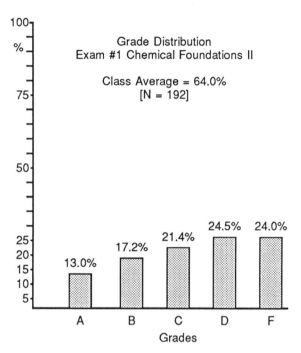

Figure 1. Bar graph, poor scale.

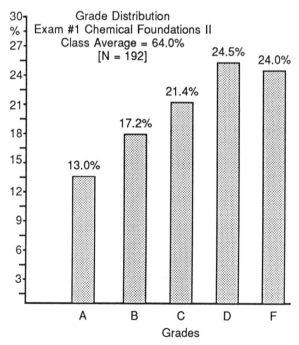

Figure 2. Bar graph, better scale.

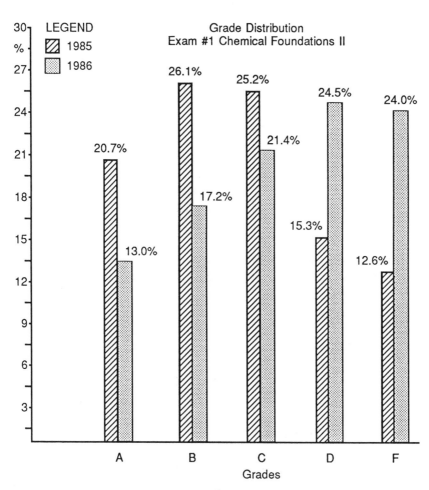

Figure 3. Bar graph comparing two years.

✔   The axes must be numbered, and the spacing on the axes should be in equal increments, but it is not necessary to number each interval. Note the differences in the following examples:

Poor

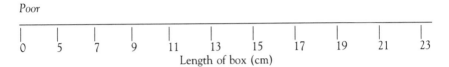

Length of box (cm)

Although the axis above is labeled well, it does not have equal increments. All the spacings increase by 2 units except for the first (from 0 to 5); if starting from 0 to 5, the increments must continue by 5 (i.e., 0, 5, 10, 15, 20, 25, etc.).

Poor

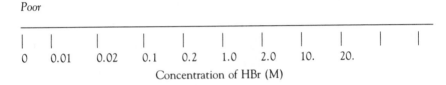

Concentration of HBr (M)

In this example, log or semilog paper should be used because a wide distribution of values must be plotted. It would not be appropriate to plot values as shown.

Better

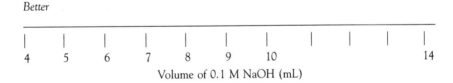

Volume of 0.1 M NaOH (mL)

This example shows equal spacing of increments; with equal spacing, not every increment needs to be labeled. The axis is appropriately labeled, and the unit of measurement is indicated. The axis does not need to begin at 0; in this situation there were no data points with values less than 4, and it would be a waste of space to include values on the axis below 4 unless it is necessary to extrapolate to a value of 0.

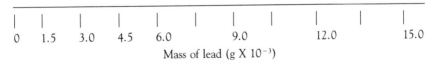

Mass of lead (g X 10^{-3})

The example just given shows numbers that would be awkward to write; they have been multiplied by a factor of 10^{-3} to bring them into a more reasonable range. The numbers actually plotted represent 1500, 3000, 4500, etc. Unfortunately, some inconsistency persists among disciplines and journals in the way in which exponential notation is interpreted. In some areas, the notation would indicate that the numbers on the axis should be multiplied by 10^{-3} when being read from the graph (i.e., the plotted numbers represent 0.0015, 0.0030, 0.0045, etc.). Usually the distinction is apparent from the subject being discussed and the manner in which other data is presented in the same journal or article.

✔ Data points on the graph should be visible. A small dot is not as easily seen as ●, ○, ×, or +. If a line is used to draw the graph, it should not obscure the data points. For example:

✔ The graph should be expanded as much as possible to fill the page, and the spacing on the axes should be chosen with this consideration in mind.

✔ Pertinent information should appear on the graph or as a caption to the graph. For example, with the Beer's law plot of $KMnO_4$, specify the wavelength at which the measurements were made, the cuvette size (or path length), and perhaps the type of spectrophotometer used.

Figure 4 shows a poorly constructed line graph. Although the axes are labeled well and the spacing is appropriate, the line should be a smooth curve and should not look like a dot-by-dot connection. When the shape of the graph is known to be a straight line or a smooth curve, it should be drawn in that manner.

✔ Multiple lines may also be plotted, usually by using different symbols for the data points, such as ●, ○, ×, etc. A key must be provided to differentiate the lines.

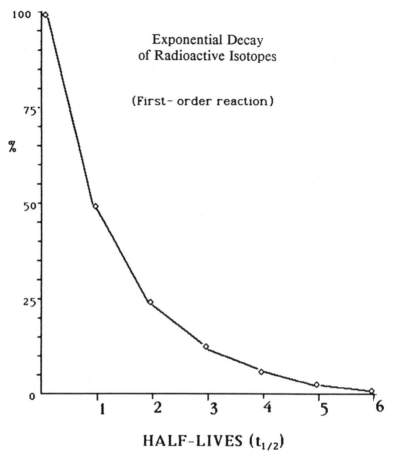

Figure 4. Poorly constructed line graph.

Charts

Charts represent another way of illustrating the presentation. A pie chart (*see* Figure 5) is primarily used when dealing with percentages that sum to 100%. In business matters, the income and expense budgets are often represented in this way. The construction of the chart is based on the mathematical equivalency that 3.6° of arc would be used to represent each 1% of the total. The first line should be drawn vertically at 12 o'clock, normally starting with the largest "piece of the pie" and decreasing as you proceed clockwise. (Figure 5 does not follow this guideline because the sequencing of the grades in traditional A–B–C order is more important.)

Three or more items should be portrayed on the pie for maximum

Grade Distribution
Exam #1 Chemical Foundations II

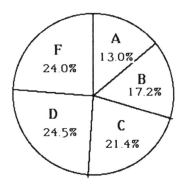

Class Average = 64.0 %
[N=192]

Figure 5. Pie chart.

effect, but when the number of items approaches eight or ten, it appears too cluttered. In that situation, combine some of the smaller items to form one piece of the pie unless they cannot logically be combined.

Flow charts are used extensively in computer science to indicate the various routes in mathematics and logic that a program follows when being used. They are effective ways to show such things as the steps in an industrial process or research project and are often used to represent different stages in planning.

A flow chart is used by almost every organization to show the hierarchy of administrative levels and the reporting lines for super-vision. Some of the biochemical cycles in living creatures are effec-tively shown in a sequential flow chart that illustrates the interactions between the various chemicals in each step.

Figure 6 shows a flow chart that illustrates several possible choices of a minor concentration in chemistry for a nonchemistry major. This flow chart could be used to advise students in a very quick way about available options.

Diagrams

Diagrams are another good way to portray scientific information, es-pecially when related to equipment or mechanical aspects of a report.

Chemistry Minors for Nonchemistry Majors

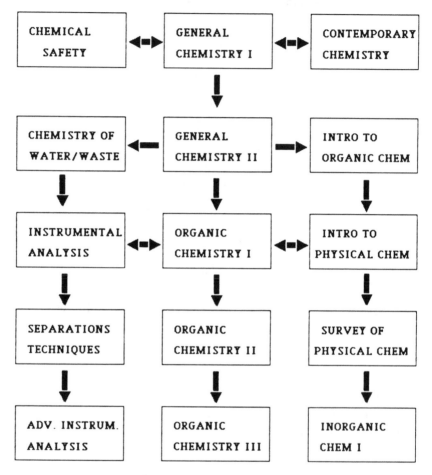

Figure 6. Typical flow chart.

As with all graphic presentations, the material must be presented clearly, sometimes by enlarging one section of a diagram so that the desired areas can be seen more easily. Anyone who has ever bought an "easy-to-assemble" toy for a child knows full well the frustration that can occur from trying to follow unclear and confusing diagrams.

Photographs

Sometimes the best way to inform the reader is with a photograph. Anyone who has seen a picture of an atomic bomb explosion is unlikely to forget it, whether he or she understands radioactivity, fission, or thermodynamics. A forest devastated by acid rain or a fire; microorganisms as seen through a microscope; changes of color in a reaction; clinical manifestations of a disease; a before-and-after sequence of a reaction; all these types of scientific phenomena are depicted well in a report with photographs.

Photographs, when compared to the other graphical representations, do have the drawback of expense in terms of reproduction costs. Many journals accept photographs only as the exception; others are eager for them. Reports that are created for in-house use often are photocopied, and many photos do not reproduce effectively; if distribution is not too large, it would be worthwhile to have duplicates of the photos made and attached to the individual reports.

Gimmicks

A graphic representation should be neat, eye-catching, and easy to interpret, and it should represent the best way to convey the meaning of the data to the reader. Newsmagazines and newspapers are usually quite inventive in finding interesting ways to show data to the readers. Professional artists can draw graphs superimposed upon brands of soda cans if they are comparing the business volume of carbonated beverage companies, or they can embellish a graph of pH versus rainfall by making the points in the shape of raindrops.

These imaginative ways of showing data are now more possible for all of us because of the availability of computer graphics. All we need is the idea. A word of caution, however: A graphic can become so artsy and so busy that the message it was designed to convey is lost. In the application of the raindrops just mentioned, to make the effect noticeable, the size of the raindrops was quite large compared to the scale of the graph. The result was a raindrop placed so that it could have represented any value between 2.4 and 3.0 inches of rain because of its elongated shape. Because the graph scale only went to a maximum of 10 inches, the uncertainty of the data point was quite large.

The use of patterns on graphs, especially bar graphs and pie charts, has been made easier with the computer. Dozens of patterns are available, and the user can create more if he or she is inventive. In Figure 3, two relatively simple patterns are used to differentiate between 1985 and 1986 data; one pattern and one plain would have been sufficient. Figure 7 shows the same pie chart as Figure 5, but

Grade Distribution
Exam #1 Chemical Foundations II

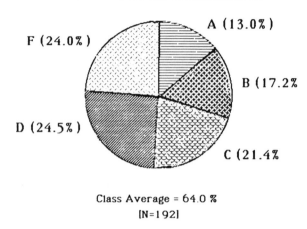

Class Average = 64.0 %
[N=192]

Figure 7. The misuse of computer graphics and design.

now each of the five areas has a pattern associated with it. Each area is fighting for the concentration of the viewer; is one area more important than the other? Figure 7 is an example of what Edward R. Tufte, in his very interesting book *The Visual Display of Quantitative Information* (1), calls "chartjunk". Instead of looking at the overall meaning of the chart (half the class had status of D or F), most viewers would wonder what special meanings were implied by the different designs. The plain pie chart in Figure 5 is obviously superior.

The other area of concern is that relative sizes of plotted material should be in proportion to changes being represented. For example, if you were trying to represent the volume of a balloon before and after heating, you might try to draw spheres to represent the two conditions. Maybe the volume increased by a factor of 2 after heating; many people would represent this effect on paper by a circle with a radius of 1 inch and a circle with a radius of 2 inches. This representation is actually misleading because the volume of a sphere is a function of the radius cubed; that means that the larger circle really depicts eight times the volume of the smaller, not twice the volume as intended.

Gimmicks and clever graphics enhance portrayal of scientific information; however, using them too frequently can draw attention away from the point that you want to make. The editors and referees of publications will certainly bring problems of this nature to the author's attention.

Reference

1. Tufte, E. R. *The Visual Display of Quantitative Information*; Graphics Press: Cheshire, CT, 1983.

Exercises

1. Arrange a visual aid presentation of the following data and discuss why you selected that method:

	Absorbance			
Wavelength (nm)	Unknown	Yellow	Red	Blue
350	0.48	0.74	0.26	0.19
375	0.76	1.30	0.29	0.21
400	1.10	1.82	0.40	0.29
425	1.26	2.00	0.46	0.19
450	1.20	1.77	0.60	0.102
475	1.10	1.24	1.03	0.130
500	1.00	0.43	1.40	0.160
525	1.03	0.18	1.52	0.220
550	0.48	0.070	0.74	0.325
575	0.160	0.050	0.180	0.61
600	0.14	0.00	0.085	0.97
625	0.23	0.00	0.100	1.70
650	0.140	0.00	0.076	1.00
675	0.070	0.00	0.060	0.17
700	0.060	0.00	0.060	0.068

2. Select a graphic presentation from a newsmagazine or newspaper and critique its effectiveness. How would you present the same data? Do it!

| Chapter 9 |

Preparing an Abstract

All manuscripts must be accompanied by an abstract. The abstracts, in general, are used directly in Chemical Abstracts (CA *indexes are prepared from the full paper). The abstracts for notes and communications will not be printed in* The Journal of Organic Chemistry; *they should be submitted on a separate sheet for direct transmittal to* Chemical Abstracts *by this journal.*

An abstract should state briefly the purpose of the research (if this is not contained in the title), the principal results, and major conclusions. Reference to structural formulas or tables in the text, by number, may be made in the abstract. For a typical paper, an 80–200 word abstract is usually adequate. (1)

An abstract is required by most professional journals as part of the submitted publication, and an abstract is frequently requested by conference organizers as part of the proposal to give an oral presentation. The two occasions differ in that an abstract for a written paper is generally composed after the text is completed, but the abstract for a talk is usually generated months before the actual event. Both have the same purpose: to allow the reader to assess the basic content quickly. After doing so, the reader can decide if he or she wishes to read the entire publication or to attend the lecture.

In many ways the abstract is the most important part of the paper or oral presentation because many more people read the abstract than actually read the paper or attend the talk. Because an abstract has this "clearinghouse" function, it must be written well, and all pertinent information must be included. Because no one could read all of the scientific literature or attend all of the scientific conferences, the abstract must be as complete as possible while remaining brief and to the point.

The terms "abstract" and "summary" are often interchanged, although they are distinctly different. A *summary* for a paper or talk assumes that a person has read the entire paper or listened to the total presentation. Only the important conclusions must be identified, and suggestions or recommendations for future activities must be made. The experimental procedure need not be reviewed, and the methodology need not be described. A summary or conclusion cannot be considered independent of the paper or talk.

An *abstract* is designed to stand alone. Its purpose is to condense the paper or talk; principal objectives, experimental methods, solution of the problem(s), and results need to be included.

As specified in the "Notice to Authors" excerpt given at the beginning of this chapter, the abstract should be short, 80–200 words (250 in some journals). At first glance, many people are thrilled that they have to write only that much but, after they have started, they realize that it is quite difficult to limit themselves to that length and present all the information that they believe is important. For this reason, an abstract needs to be written, rewritten, and generally rewritten again. You can minimize the number of words by deleting some of the superfluous phrases that we use automatically (see Chapter 2).

Types of Abstracts

Two types of abstracts are used. The *indicative* or *descriptive abstract* outlines the development of the paper or talk. In many ways, it might resemble a table of contents. Normally, it is used for review papers, conference reports, governmental reports, and books. It is often used for oral presentations because frequently the proposal is made before the specific experimental results are completed.

The *informative abstract* capsulizes the paper or talk and provides the results and other important points of the presentation. It is most frequently used for journal articles and would be similar to the one shown in Figure 1 (2). The following information should be given:

1. Principal objectives: What was the purpose of the experiment? Why was it done? What was the expected conclusion? In the

abstract in Figure 1, the stated purpose was to determine the eutrophication potential of Lake Mead.

2. Methodology: How was the investigation carried out? Experimental details should not be given. For example, in Figure 1, an analysis was made using EDTA. To find the details, such as molarity, temperature, or extraction solvent, you would need to go directly to the primary source.

3. Problems encountered, if any: What unexpected situations or difficulties emerged? In the abstract shown, filtered samples were used and these were not equivalent to what was observed in the lake, so the investigators tried another step, autoclaving, to improve the compatibility.

4. Results and conclusions: Were the objectives achieved? Is this the end of this project or are there future investigations? In the example, the investigators identified the limiting nutrients, P and N, and characterized the productivity potential for Las Vegas Bay and Boulder Basin.

Chemical nomenclature mentioned in the abstract should follow the standard systematic IUPAC guidelines, although a trivial name can

---

105: **84750s The effect of secondary effluents on eutrophication in Las Vegas Bay, Lake Mead, Nevada.** Greene, Joseph C.; Miller, W. E.; Merwin, Ellen (Corvallis Environ. Res. Lab., U.S. Environ. Prot. Agency, Corvallis, OR 97333 USA). *Water, Air, Soil Pollut.* **1986**, *291(4)*, 391–402 (Eng). The eutrophication potential of Lake Mead, with primary emphasis on Las Vegas Bay, Nevada, was detd. using *Selenastrum capricornutum*. Nutrient limitation profiles were detd. for 3 sampling stations in Las Vegas Bay and 2 in Boulder Basin. After heavy metals were chelated with EDTA, P was identified as the primary limiting nutrient, with N being the secondary limiting nutrient for *S. capricornutum*. Productivity potential was highest in upper Las Vegas Bay near the sewage inflow. Toward the mouth of the bay and in Boulder Basin, progressively lower potentials were defined. Productivity potential could not be predicted from the filtered samples because the nutrients bound up in the indigenous biomass remained on the filters. Autoclaving followed by filtration prior to assay enabled the algae to produce yields relative to the productivity obsd. in the lake.

Figure 1. A typical abstract. (Reproduced from Ref. 2. Copyright 1986 American Chemical Society.)

be used if it is well-established. Abbreviations should be defined when first introduced unless they are considered routine (e.g., in the sample abstract, EDTA is not defined because it is accepted as meaning ethylenediaminetetraacetic acid). Some journals will not permit tables and graphs to be cited in the abstract, but others allow this practice.

Literature references should not be included in the abstract unless they are an integral part of the experiment. An exception would be if someone revised or adapted a previously published experimental methodology and needed to refer to it continually. Structures of compounds and some equations are often included in the abstract, especially if they clarify the nature of the work being presented. The primary deciding factor for all of these matters is whether inclusion of the information is necessary for the reader to fully understand the scope of the publication or presentation.

As mentioned in the excerpt from *The Journal of Organic Chemistry*'s "Notice to Authors", most abstracts are presented directly in *Chemical Abstracts* even if the abstract is not included with the published paper. This practice is true for other scientific areas; for example, *Biological Abstracts, Solid State Abstracts, Nuclear Science Abstracts, Physics Abstracts,* and many more in specific disciplines. Material that does not have an abstract provided by the authors is read and condensed by professional writers employed by the various abstracting services.

It is important to have an informative title and abstract because one form of indexing uses key words. Figure 1 can be found in the key word subject index (of *Chemical Abstracts*) under

> Eutrophication
> Las Vegas Bay Nevada 84750s

> *Selenastrum*
> *capricornutum* lake eutrophication Nevada 84750s

and

> lake
> eutrophication *Selenastrum capricornutum* Nevada 84750s

The article was not listed under any of the following possible key words:

algae	Las Vegas
bay	Nevada
Boulder	nutrient
capricornutum	potential
EDTA	productivity
effluent	secondary

Although *Chemical Abstracts* and many other abstracting services take .ey words from the entire article, others, such as *Chemical Titles*, rely only upon the title for indexing. For example, the article entitled "Bioorganic studies of visual pigments. 4. A mechanistic model study of processes in the vertebrate and invertebrate visual cycles" (3) is listed in the July 18, 1986, issue of *Chemical Titles* (No. 15) under the following (boldface is added for emphasis):

A mechanistic model stud + **Bioorganic** studies of visual pigments. 4.
brate and invertebrate visual **cycles.** = + y of processes in the verte
ocesses in the vertebrate and **invertebrate** visual cycles. = + y of pr
Bioorganic studies of visual **pigments.** 4. A mechanistic model study el
study of processes in the **vertebrate** and invertebrate visual cycles
e vertebrate and invertebrate **visual** cycles. = + y of processes in th
stud + Bioorganic studies of **visual** pigments. 4. A mechanistic model

Writing the Abstract

Figure 2 contains the text of a communication to the editor of *Inorganic Chemistry* reporting the important and chemically significant finding that fluorine (F_2) can be synthesized (4). It is brief (about 600 words), as is appropriate for a communication, and does not have an abstract published with it. (Notes and communications generally do not have abstracts published in the journal, although the author submits one to be sent to *Chemical Abstracts*.) We will use this short text (Figure 2) to examine the approach to preparing an abstract.

1. Read the entire article two or three times so that you completely understand the information being presented.
2. Underline the information that appears to be most important. Omit most background, history, experimental details, opin-

Chemical Synthesis of Elemental Fluorine

The chemical synthesis[1] of elemental fluorine has been pursued for at least 173 years[2] by many notable chemists, including Davy[2], Fremy[3], Moissan[4], and Ruff.[5] All their attempts have failed, and the only known practical synthesis of F_2 is Moissan's electrochemical process, which was discovered exactly 100 years ago.[6]
　　Although in principle the thermal decomposition of any fluoride is bound to yield fluorine, the required reaction temperatures and conditions are so extreme that rapid reaction of the evolved fluorine with the hot reactor walls preempts the isolation of significant amounts of fluorine. Thus, even in the well-publicized case of K_3PbF_7,[7,8] only trace amounts of fluorine were isolated.[5,9]
　　These failures, combined with the fact that fluorine is the most electronegative element and generally exhibits the highest single bond energies in its combinations with other elements[10], have led to the widely accepted[11-15] belief that it is impossible to generate fluorine by purely chemical means.
　　The purpose of this communication is to report the first purely chemical synthesis of elemental fluorine in significant yield and concentration. This synthesis is based on the fact that thermodynamically unstable high-oxidation-state transition-metal fluorides can be stabilized by anion formation. Thus, unstable NiF_4, CuF_4, or MnF_4 can be stabilized in the form of their corresponding MF_6^{2-} anions. Furthermore, it is well-known that a weaker Lewis acid, such as MF_4, can be displaced from its salts by a stronger Lewis acid, such as SbF_5.

Side annotations:
Omit background

Combine as introductory sentence

Part of introductory material

Include as factual material

Exclude examples

Figure 2. Article for abstracting. Footnotes are indicated but references have been omitted. (Reproduced from Ref. 4. Copyright 1986 American Chemical Society.)

$$K_2MF_6 + 2\ SbF_5 \rightarrow 2\ KSbF_6 + [MF_4] \quad (1)$$

If the liberated MF_4 is thermodynamically unstable, it will spontaneously decompose to a lower fluoride, such as MF_3 or MF_2, with simultaneous evolution of elemental fluorine.

Delete general information

$$[MF_4] \rightarrow MF_3 + \tfrac{1}{2}\ F_2 \quad (2)$$

Since a reversal of (2) is thermodynamically not favored, fluorine can be generated even at relatively high pressures.

Consequently, the chemical generation of elemental fluorine might be accomplished by a very simple displacement reaction, provided a suitable complex fluoro anion is selected which can be prepared without the use of elemental fluorine and is derived from a thermodynamically unstable parent molecule. The salt selected for this study was K_2MnF_6. It has been known[16] since 1899 and is best prepared from aqueous solution.[17]

Include rationale

Delete historical details and preparation

$$2\ KMnO_4 + 2\ KF + 10\ HF + 3\ H_2O_2$$

$$\xrightarrow{50\%\ aq\ HF} 2\ K_2MnF_6 + 8\ H_2O + 3\ O_2 \quad (3)$$

The literature yield of 30% was increased to 73% and can probably be improved further by refining the washing procedure (use of acetone instead of HF)[18]. The other starting material, SbF_5, can be prepared[19] in high yield from $SbCl_5$ and HF.

Keep significant details

$$SbCl_5 + 5\ HF \rightarrow SbF_5 + 5\ HCl \quad (4)$$

Figure 2.—Continued. *Article for abstracting.* Continued on next page.

Since both starting materials, K_2MnF_6 and SbF$_5$, can be readily prepared without the use of F$_2$ from HF solutions, the reaction

$$K_2MnF_6 + 2\ SbF_5 \rightarrow$$
$$2\ KSbF_6 + MnF_3 + \tfrac{1}{2}\ F_2 \qquad (5)$$

Keep vital information

represents a truly chemical synthesis of elemental fluorine.

The displacement reaction between K_2MnF_6 and SbF$_5$ was carried out in a passivated Teflon-stainless-steel reactor at 150 °C for 1 h. The gas, volatile at -196 °C, was measured by PVT and shown by its reaction with mercury and its characteristic odor to be fluorine. The yield of fluorine based on (5) was found to be reproducible and in excess of 40% but most likely can be improved upon significantly by refinement of the experimental conditions. Fluorine pressures of more than 1 atm were generated in this manner.

Keep important results but omit author's opinion

In summary, the purely chemical generation of elemental fluorine can be achieved in high yield and concentration by a very simple displacement reaction between starting materials that can be prepared in high yields from HF solutions and have been known for 80 years or longer. As in the cases of noble gas[20] or NF$_4$[21] chemistry, the successful synthesis of elemental fluorine demonstrates that one should never cease to critically challenge accepted dogmas.

Omit repetition of facts

Omit author's opinion

Figure 2.—Continued.

ions, examples, imprecise data, repetitions, and obvious explanations. See Figure 2 for an example of selecting the key material.

3. Without worrying about sentence structure or grammar, reproduce the underlined material.

> . . .the only known practical synthesis of F_2 is Moissan's electrochemical process . . . although in principle the thermal decomposition of any fluoride is bound to yield fluorine . . . failures . . . fluorine is the most electronegative element . . . highest single bond energies . . . impossible to generate fluorine by purely chemical means . . . report the first purely chemical synthesis of elemental fluorine in significant yield and concentration . . . thermodynamically unstable high-oxidation-state transition-metal fluorides can be stabilized by anion formation . . . suitable complex fluoro anion . . . prepared without the use of elemental fluorine . . . thermodynamically unstable parent molecule . . . K_2MnF_6. . . yield of 30% was increased to 73% . . . SbF_5 can be prepared in high yield from $SbCl_5$ and HF . . . K_2MnF_6 and SbF_5 . . . prepared without the use of F_2 . . . K_2MnF_6 + 2 SbF_5 → 2 $KSbF_6$ + MnF_3 + ½ F_2 (5) . . . was measured by PVT. . . shown by its reaction with mercury . . . the yield of fluorine based on (5) was found to be reproducible and in excess of 40% . . . fluorine pressures of more than 1 atm were generated in this manner.

This process has resulted in reducing the material to about 25% of its original length.

4. Rewrite the underlined information in your own words. Make complete sentences, but concentrate more on content than on style.

Elemental fluorine (F_2) has only been synthesized previously through electrochemical methods. Several attempts have been made at chemical synthesis, but it is generally accepted that this is not possible because of fluorine's high electronegativity and high single-bond energies. The author states that, because unstable high-oxidation-state transition-metal fluorides can be stabilized by anion formation and that a weaker Lewis acid can be replaced by a stronger Lewis acid, K_2MnF_6 and SbF_5, respectively, were chosen as possible candidates to produce

F_2. Both can be prepared in high yield without the use of elemental fluorine, using HF instead. The yield of K_2MnF_6 was increased to 73% and could be further improved. When the two compounds were reacted, K_2MnF_6 + 2 SbF_5 → 2 $KSbF_6$ + MnF_3 + ½ F_2 (5), F_2 was produced (40% +) yield), giving pressures of more than 1 atm. It was characterized by PVT data and reaction with mercury.

This step has not condensed the material much, but presents it in whole sentences, although some are rather awkwardly constructed.

5. If possible, the abstract should be further reduced. Unnecessary phrases should be removed, sentences combined, repetitious material deleted, and the final condensation polished.

Elemental fluorine (F_2) has been chemically synthesized, although, until now, only electrochemically. Chemical methods were thought to be impossible because of fluorine's high electronegativity and single-bond energies. K_2MnF_6, a high-oxidation-state transition-metal fluoride and a weak Lewis acid, was prepared with an increased yield (73%) using KF and HF. SbF_5, a stronger Lewis acid, was also prepared from HF, ensuring no contamination from elemental fluorine. When reacted, K_2MnF_6 + 2 SbF_5 → 2 $KSbF_6$ + MnF_3 + ½ F_2. F_2 was produced in 40% + yield at pressures greater than 1 atm. It was positively characterized through PVT and reaction with mercury.

Now there has been a substantial reduction to about 15% of the original material.

6. Check your abstract against the original article. Can a person reading the abstract learn what was accomplished? How was it done (in general terms)? What was its significance? If you can get the answers to these questions from your condensation, you have successfully prepared an informative abstract.

The actual abstract as it appeared in *Chemical Abstracts* (5) is much more condensed than the version just given but contains all of the essential information:

105: **163745x Chemical synthesis of elemental fluorine.** Christe, Karl O. (Rocketdyne Div., Rockwell Int., Canoga Park, CA 91303 USA). *Inorg. Chem.* **1986,** *25(21)*, 3721–2 (Eng).

After 173 yr of futile efforts the 1st chem. prepn. of elemental F in significant yield and concn. has been achieved. The prepn. is based on the reaction $K_2MnF_6 + 2\ SbF_5 \rightarrow 2\ KSbF_6 + MnF_3 + 0.5\ F_2$ at 150 °C, and both starting materials can readily be prepd. from HF soln.

If you were preparing an indicative or descriptive abstract, it could be even shorter than the one published because most details would be omitted. For example, for the same article, the indicative abstract might be

The first successful chemical (nonelectrochemical) synthesis of elemental fluorine (F_2) through a Lewis acid displacement reaction between K_2MnF_6 and SbF_5 is documented.

This abstract gives a good indication about what information is contained in the article, but a person would really have to go directly to the source if he or she were interested in even the most general details.

References

1. *J. Org. Chem.* **1986,** 51, 9A.
2. *Chem. Abstr.* **1986,** 105(10), 379.
3. Liu, R. S.; Matsumoto, H; Asato, A. E.; Mead, D. *J. Am. Chem. Soc.* **1986,** 108, 3796–3799.
4. Christe, K. O. *Inorg. Chem.* **1986,** 25, 3721–3722.
5. *Chem. Abstr.* **1986,** 105(18), 736.

Exercises

1. Select a journal article that has an abstract and classify it as indicative or informative. Specify how it presents (or does not present) necessary information according to the four guidelines in the text.

2. Rewrite the abstract used in Exercise 1 and try to decrease its length.

3. Select an article from the science section of a current newsmagazine (e.g., *TIME, Newsweek*) and write an abstract for it. Make it as brief as possible without omitting important information.

| Chapter 10 |

Using Computers and Word Processors

The availability of computers, software packages, and word processing equipment has increased dramatically in the past 5 to 10 years. Most of the commercially available materials are user-friendly and relatively easy to manipulate. These timesaving devices certainly make writing and revising much easier tasks.

As with any new development, the user must realize the limitations of the equipment and be concerned with understanding the complete functioning of the model. Revisions and new applications of software are emerging almost daily, so that uses are available today that did not exist at the time of writing this chapter. Software and hardware packages exist in such variety and proliferate so quickly that it is not practical to describe any in specific detail in this text.

Many of the software packages are similar and can be described in general terms. You would be wise to seek specific information from the descriptive booklet accompanying the model to which you have access. (The booklet is, I hope, a good example of technical communication!)

Changes in Text

Word processors allow you to compose written material at a keyboard and have it appear on a monitor. Many new users have difficulty creating text at the keyboard and may have to write it on paper, even

1451–4/88/0101$06.00/1 © 1988 American Chemical Society

roughly, before entering it into the processor. The advantage either way is that, once the text is entered, it can be changed readily.

Commands allow you to insert or delete a letter, word, sentence, paragraph, or page. Usually, these commands are invoked by positioning a cursor (pointer) where the change is desired and then "calling" a command (by pressing a key) such as "add" or "insert", "delete" or "remove", and "change" or "edit". Complete areas of text can be removed permanently, or they can be stored temporarily and moved to another location with functions called "cut and paste" or "move". A similar function called "copy" or "pick and put" enables you to take a part of the text and place it in other areas without removing the original section from the text. This feature is especially useful if a complex phrase or equation is to be used repeatedly.

Other editing commands include the ability to "search" for misspelled words or phrases and "replace" them with the correctly spelled versions. Usually you are given the option of replacing each one as you locate it. For example, if you realize that you have consistently typed "mispelled" instead of "misspelled" and have used that word frequently throughout the paper, it is much easier to have the processor search for it than to proofread the entire text yourself. This option allows you to change your mind and replace "ACS" with "American Chemical Society" throughout, if desired. Caution must be advised on the use of the global "replace" command. In some systems, if you wanted to change the word "out" to "outing", the "out" combination would be changed everywhere, and "throughout" would become "throughouting" and the phrase "down and out" would then be "down and outing"! Most word processors have a "query replace", which allows you to see each occurrence of the word and replace the word or go on to the next occurrence.

Another common feature is a command that allows you to "undo" the most recent change, if you decide the original was better. Many versions enable you to "goto" a specific page, sentence, or paragraph. This feature is particularly useful if you have only one change (for example, on page 14) and do not want to go through the entire text to get to that point.

Text Format

All processors have the basic features of typewriters and allow you to specify the width of left and right margins, the number of spaces to be indented at paragraphs, and the location of columns in tabulation. In addition, most processors enable you to specify the amount of empty space to be allotted at the top (headers) and bottom (footers) of the page. Usually, built-in default values are followed if you do nothing to change them. Alignment of text along the left and right margins is flexible with most software packages. The default option normally indicates alignment along the left margin as is the case with manually typed material. Some versions permit alignment at the right margin only

as is shown with this portion of text.

The most frequently used alignment, or *justification*, places the text flush with both margins as is normally seen with printed or typeset material and with this sentence.

Another frequent choice is to have the material centered,
which is usually selected for titles
(and here in this sentence).

Most of these options can be altered throughout the text by selecting the appropriate command. You have to be careful, however, because some packages have the feature that if you change a margin or tab in the beginning of the text, it becomes changed throughout. This change is not always what you want.

Justified text may not always be best in appearance. Software that does not have automatic hyphenation capabilities will occasionally move a long word to the next line and spread out the words on the previous line so that the spacing between words becomes large. In some cases, the minimum and maximum amounts of space between words can be specified by the user. A colleague of mine recently wrote a memo that was very difficult to read because of this spacing problem. It looked like this:

I have become aware   of the   fact   that the parking lot
is not being plowed on   Sunday evenings . This causes problems   , because   I usually work   late   on   those
days to prepare   for Mondays.

Type Sizes, Faces, and Fonts

Many word processors allow the user to vary the style of the type being used. Some common choices are

- plain text
- **boldface**
- *italic*
- underlining
- outlines
- shadow

and combinations of those choices such as bold and italic or underlined and italic. In addition, many allow the type size to vary:

- 10 point
- 12 point
- 24 point

Some software packages have a selection of typefaces to choose from as well. Typical examples are

Chicago	Monaco
Courier	New York
Geneva	Times
Helvetica	Venice

Caution is advised on mixing too many of these typefaces in one document. On seminar announcements or similar flyers where the designer can become a bit artistic, keep in mind that the message is much more easily read (and after all, you want the message to be read) if only one or two type sizes, typefaces, and type fonts are used.

Drawing

In addition to word processing capabilities, many computers have software packages that can be used for creating drawings and other graphics. These packages vary greatly in flexibility of commands, and you will need to read the specific directions for the available setup. Depending on the system, these packages can be used to design company logos; to illustrate laboratory equipment; to do freehand drawing; or to create letterheads, memo headings, or borders. These embellishments can help supplement technical writing.

Hardware

Hardware varies as greatly as the software packages, and new features are introduced almost as quickly. The tremendous advantage of virtually all systems is the ability to store, retrieve, and alter material indefinitely, usually on disks or diskettes, hard or floppy. In addition, many small computers have the capacity to communicate with large main-frame computers and can use their storage capacity.

For technical writing, a printer is a necessary component of the system. For most professional uses, it should be a letter-quality printer (i.e., the type looks as if it were typewritten) and not the dot-matrix style (the type looks like a series of connected dots). However, vast improvements have occurred in the quality of dot-matrix printers and the letters on a good dot-matrix printer do not look as rough as they once did. Laser printers are available in a wide range of models and produce very good copy. Not all printers are capable of producing graphics, so you must ensure that the software and printer are compatible.

Computer-Generated Manuscripts

It is possible to generate a text or manuscript on a computer or word processor, store it on a disk, and mail it directly to the publisher where it can be machine-read. Using a modem, it is possible to transmit the material via telecommunications. However, the publisher must have systems that are compatible with the author's; with the wide variety of software and hardware available, compatibility is often a problem. Many publishers have the ability to convert diskettes from one system to a form compatible with their system.

Another difficulty is that to date no uniform coding exists for the transmission and preparation of the manuscript, although the American Chemical Society has established guidelines for its publications (1). At first glance, the procedures look difficult and awkward, but once learned, they could provide a more efficient way of handling textual material. The process should shorten the time elapsed between submitting a manuscript and publication of the information. The number of publications submitted in machine-readable form could increase dramatically, especially if the special coding requirements are refined and eventually become uniform among all publishers.

Reference

1. Brogan, M. C. In *The ACS Style Guide*; Dodd, J. S., Ed.; American Chemical Society: Washington, DC, 1986; Chapter 5, pp 149–157.

| Chapter 11 |

Proofreading

If you have ever, like me,
Missed the "r" and hit the "t",
Addressing some fat blister
As "Mt." instead of "Mr.",
I trust you left it unamended?

Splendid

J. B. Boothroyd (*1*)

Although this poem treats proofreading in a humorous way, proofreading is one of the most important steps in the writing process. No good excuse has ever been found for allowing typographical errors to remain in a final copy of a paper or manuscript, and the increasing availability of word processing equipment and computers leaves even less justification for an imperfect copy.

Once the final copy is typed, how do you successfully approach the act of proofreading? Most people feel that if they read the paper once or twice, they will spot the spelling errors, typographical errors, and omitted words. This idea could not be further from the truth. People reading their own writing at a normal pace will be able to judge if a sentence is misplaced or if the flow of the paper is correct, but will miss most of the typing errors. The writer/proofreader knows what he or she intended to write and will normally read the words as such, whether the words are right or wrong.

If you are doing your own proofreading, study the manuscript by

1451–4/88/0107$06.00/1 © 1988 American Chemical Society

reading it slowly and looking at each word. One colleague of mine advises that the best way to see errors is to read the paper backward, from finish to start! Another uses an index card to cover everything except the line of type that is being read.

A somewhat easier method is to ask a friend or colleague to read the paper and help find the problem areas. An even more effective way is to do it together, with one person reading the original paper aloud while the other compares it to the typed copy.

In the sciences, proofreading that is not thorough can lead to some embarrassing, and possibly dangerous, results. The difference between "alkanes", "alkenes", and "alkynes" is simply one letter, but the difference in properties and reactivities is often drastic. A solution that is 0.71 M may react in an undesired manner if the directions for the experiment actually specified 0.17 M. If the material is intended for publication, a printer or an editor would certainly know to change "hte" to "the" but would not know which of the compounds or molarities mentioned earlier were correct.

Spelling correction programs are available for most home or office computers, and these are of immense value to people who are notoriously poor at this task. Many of these programs do not have vocabulary lists of words peculiar to the various scientific and technical areas, although most have the capability of having additional terms entered by the user. Even with that flexibility, "alkyne" would not be isolated by the program as a misspelled word, and the typist would have to look carefully at each compound's name.

Once the manuscript has been submitted, reviewed, and accepted, the author will receive galley proofs that he or she should read very carefully. The "Notice to Authors of Papers" in *Inorganic Chemistry* (2) states

> The attention of authors is called to the Instructions which accompany the proof, especially the requirement that all corrections, revisions, and additions be entered on the proof and not on the manuscript. Proofs should be checked against the manuscript (in particular all tables, equations, and formulas), as this is not done by the editor, and returned as soon as possible, for no paper is released for printing until the author's proof has been received.

As noted previously, when you are proofreading the galley copy, compare it to the original manuscript. Correct any printing errors but keep other changes to a minimum. Occasionally you need to add revised data or a recently published reference; this addition can often be done as a postscript, "Note Added in Proof". If a new reference is included, it is more convenient and less error-prone to add it at the end of the paper; otherwise many or all of the reference numbers in the text will need revising, and invariably one is overlooked. Proofreading instructions sent by the publisher or editor will inform you about the possible ways of handling these additions.

In journals, if extensive changes (from the original manuscript) are required, the paper may need to be resubmitted to the editor for approval. Often this step will require going through the review process again. If another review is not required, the changes will almost certainly delay the date of publication, and some journals will charge the author for part or all of the direct cost of resetting the type.

Checkpoints for Proofreading

✔ Compare the galley proof with the original manuscript. In addition to the text, data, and references, carefully check quality, numbering or sequencing, and captions for illustrations, tables, charts, and graphs.

✔ Scrutinize the atomic symbols, subscripts, bonds, and coefficients in formulas, equations, and structures.

✔ Compare the spelling with the original for names of compounds, names of biological species, and proper names.

✔ Do not obscure type in the proof. Make corrections on the edge of the galley and use a single line (to delete) or a caret (to insert) in the text.

Table I shows the more common markings used in correcting errors in a manuscript, with examples of their use and the corrected version.

Exercise

1. Assuming that you are not a perfect typist, select a page in a chemistry (or other) text and type it in double-spaced format without worrying about typographical errors. Using proofreading marks shown in Table I, correct the typing as if it were a manuscript.

Table I. Proofreaders' Marks and Sample Copy

Instruction	Mark in Margin	Mark in Manuscript	Corrected Version
		Punctuation	
insert a comma		use benzene but	use benzene, but
insert an apostrophe		Id rather go now	I'd rather go now
quotation marks		but he said, No!	but he said, "No!"
period		the end Then	the end. Then
question mark		couldn't he I	couldn't he? I
colon		the following ideas	the following ideas:
semicolon		Rochester, NY Buffalo	Rochester, NY; Buffalo
hyphen		solid state physics	solid-state physics
parentheses		polyethylene is	poly(ethylene) is
brackets		bicyclo 2.2.1 hept-5-en-2-yl	bicyclo[2.2.1]hept-5-en-2-yl
		Spacing	
delete		delete ~~delete~~ the word	delete the word
delete and close		delete the letter	delete the letter
close		close the space	close the space
insert		insert a space	insert a space
new paragraph		and it should. But the next paragraph	and it should. / But the next paragraph

Instruction	Mark	Marked copy	Corrected result
no paragraph		and it should. But if you don't know	and it should. But if you don't know
		Alignment	*Alignment*
center		center / usually with a title	center / usually with a title
move up		and it moved	and it moved
move down		and it moved	and it moved
align horizontally		and the alkanes	and the alkanes
align vertically		when a list / needs to be / aligned	when a list / needs to be / aligned
subscript		H_2O is the best	H_2O is the best
superscript		Na is an ion	Na^+ is an ion
		Type font	*Type font*
lower case		sometimes in the	sometimes in the
capitalize		Albert einstein is	Albert Einstein is
small caps		β-d-glucose	β-D-glucose
italics		to define peptides	to define *peptides*
roman		to change back to	to change back to
boldface		in a manuscript	in a **manuscript**
wrong font		to change to	to change to

References

1. *The Silver Treasury of Light Verse*; Williams, O., Ed.; The New American Library: New York, 1957; "Please Excuse Typing", p 313.
2. *Inorg. Chem.* **1986,** *25,* 6A.

SPECIFIC NEEDS

| Chapter 12 |

Academic Laboratory Reports

Every student has to write a laboratory report in some scientific area; if majoring in the sciences or engineering, the student is likely to think that these reports are much too numerous and demanding. The task will, however, become less demanding and time consuming if the skill of assembling a good laboratory report is acquired early in the student's academic career. It also will be a definite asset in attaining success in professional life, whether the chosen occupation is academic, industrial, or research.

The key to success is maintaining a complete laboratory notebook (*see* Chapter 8 and also *Writing the Laboratory Notebook* [1]). If the information and data needed are easily found in the notebook, compiling the report will be less time consuming.

Although the format required for an academic laboratory report may vary from one subject to another or from one instructor to another, all laboratory reports have similar contents. Typical sections are

- Purpose (or Objective)
- Procedure
- Data
- Sample Calculations
- Results
- Conclusion(s)
- Interpretations or Recommendations

1451–4/88/0115$06.00/1 © 1988 American Chemical Society

Additional sections might include

- Theory
- Background Material
- Equipment
- Chemicals List
- Diagram of Apparatus
- Diagram of Results
- Error Analysis

and others specified by the instructor.

Purpose (or Objective)

For some reason, many students find the purpose or objective of the experiment to be the most difficult part of the laboratory report to write. If you truly understand the nature of the experiment or project, the purpose should be the easiest statement to make and you should be able to state it before beginning the actual task. It ought to answer all of the following questions:

- Why are you doing this? (In addition to the fact that you must do it to pass the course or keep your job!)
- What do you hope to find (prove, make, show)?
- What principle will be illustrated?

The statement of the objective is one part of the laboratory report that does not have to be a complete sentence, unless full sentences are required by the instructor. The following examples contrast a poor statement of purpose and a better one:

Poor

Purpose: To find the answer for an unknown water sample.

Better

Purpose: To analyze an unknown water sample for chemical oxygen demand (COD) by the dichromate method.

Poor

Purpose: Magnesium and oxygen will be reacted and analyzed.

Better

Purpose: Magnesium will be burned in air, and the product(s) will be analyzed gravimetrically to determine the empirical formula(s).

Procedure(s)

The amount of descriptive material in the Procedures section may vary according to the instructor's directions. If a standard procedure is used, giving it a reference and stating any alterations may be sufficient. For the experiment involving the reaction of magnesium with oxygen, this section could be

> Procedure: The method followed is given in Experiment 13, Section A, of *Laboratory Experiments: Basic Chemistry* (4th Ed.) by Charles H. Corwin, Prentice-Hall: Englewood Cliffs, NJ, 1985; pp 131–134.

If a minor change in the procedure was made, the following statement could be made:

> Procedure: The method followed is given in Experiment 13, Section A, of *Laboratory Experiments: Basic Chemistry* (4th Ed.) by Charles H. Corwin, Prentice-Hall: Englewood Cliffs, NJ, 1985; pp 131–134. Three determinations were performed instead of the two designated in the text.

On other occasions, it may be desirable to reference the procedure and summarize it in your words, omitting the drop-by-drop explanation of all measurements and chemicals used. Using the same example:

> Procedure: The method followed is given in Experiment 13, Section A, of *Laboratory Experiments: Basic Chemistry* (4th Ed.) by Charles H. Corwin, Prentice-Hall: Englewood Cliffs, NJ, 1985; pp 131–134.

A porcelain crucible was cleaned, heated, cooled, and weighed. The crucible was reweighed containing a piece of magnesium ribbon and was heated until the magnesium began to burn. After complete reaction, the crucible was cooled and water was added. The crucible was reheated, cooled, and reweighed. The empirical formula of the resulting magesium oxide was calculated.

For a formal laboratory report, reference the method you used (if it is not original) and be very specific about what you did so that the reader could repeat the experiment exactly by looking at the report only. Your description should paraphrase the directions; it should not be a direct quote. For example:

Procedure: The method followed is given in Experiment 13, Section A, of *Laboratory Experiments: Basic Chemistry* (4th Ed.) by Charles H. Corwin, Prentice-Hall: Englewood Cliffs, NJ, 1985; pp 131–134.

1. Heat a clean crucible and cover (on a clay triangle) to red heat; cool and weigh.

2. Form a 25-cm magnesium ribbon into a coil and place in the crucible. Reweigh with the cover on.

3. With the cover off, heat the crucible to red heat until the magnesium begins to spark. Replace the cover and continue heating until there is no sign of reaction when the cover is carefully removed.

4. Cool the crucible to room temperature and add several drops of distilled water; fire to red heat for 5 minutes; let cool to room temperature and weigh.

5. Calculate the empirical formula for the compound formed between magnesium and oxygen.

Often, you will need to include a diagram of the experimental setup to clarify the procedure. If a special piece of equipment is used, include a description or diagram. If unusual chemicals are required, mention how they can be obtained. Point out any safety hazards in the procedure here or in a special section designated "Safety".

Data

Tabular Data

Enter all measurements and observations made during the experiment into the laboratory notebook in an organized manner. Usually, it is unnecessary to reproduce all of the rough measurements in the Data section of the report. For example, the laboratory notebook might contain the following entries as raw data

Data:
1. Mass of crucible   24.5681 g
2. Mass of crucible + magnesium
     (before heating)   25.7793 g
3. Mass of crucible (after heating)   26.5759 g

The important information in this experiment is not the mass of the crucible or of the crucible and magnesium, but the mass of the magnesium itself. The data presented in the report would represent the differences in masses and would reference the original or raw data:

Data:[*]
1. Mass of magnesium   1.2112 g
2. Mass of compound formed between
     magnesium and oxygen   2.0078 g
3. Mass of oxygen in compound   0.7966 g

[*]Original data is on page 73 of laboratory notebook.

Often several trials are performed using the same procedure. In this situation, a table is the best way to present the information:

Data:[*]

	Trial 1	Trial 2	Trial 3
1. Mass of magnesium	1.2112 g	1.2459 g	1.1443 g
2. Mass of compound formed between Mg and O	2.0078 g	2.0653 g	1.8970 g
3. Mass of oxygen in compound	0.7966 g	0.8194 g	0.7526 g

[*]Original data is on page 73 of laboratory notebook.

Units should be included with all measurements and can be placed with the number itself (as in the example just given) or they can be indicated in the description of the measurement:

Data: *

	Trial 1	Trial 2	Trial 3
1. Mass of magnesium (g)	1.2112	1.2459	1.1443

*Original data is on page 73 of laboratory notebook.

Graphical or Nontabular Data

Not all data are numerical. Spectra, chromatograms, and computer output might be attached to the report (or referred to), or a photocopy might be reproduced in the Data section. In some situations, the Data section might more appropriately be labeled "Observations", especially for nonquantitative experiments. Certain biology experiments might require a Data section that is composed of sketches labeled with the pertinent observations or peculiarities.

Often the data are more easily represented in a graph than in a table; occasionally, it is convenient to use both methods. If you want to use a graph, see Chapter 8 for guidelines to prepare an effective presentation.

Sample Calculations

Many academic laboratory reports include an example of the calculation used to obtain the answer so that the reader can judge the mathematical validity of the results. If you choose one trial and show its progress through the steps of the calculation, readers assume that the remaining trials were treated in the same manner. A typical sequence is shown for Trial 1 of the data given previously:

Sample Calculations:

Moles of oxygen:

$$0.7966 \text{ g O} \times \frac{1 \text{ mole O}}{15.99 \text{ g O}} = 0.04982 \text{ moles O}$$

Moles of magnesium:

$$1.2112 \text{ g Mg} \times \frac{1 \text{ mole Mg}}{24.31 \text{ g Mg}} = 0.04982 \text{ moles Mg}$$

Empirical formula:

Mg $_{0.04982}$O$_{0.04982}$

Mg $_{\frac{0.04982}{0.04982}}$ O$_{\frac{0.04982}{0.04982}}$ or MgO

If estimates of error in the answer are required, they may appear in this section or under a separate heading (e.g., "Propagation of Errors"). Calculation of precision, standard deviations, and other statistical treatment of the data may appear with the sample calculations. Often these will be presented in the Conclusion or Discussion of Results sections.

Results

The "Results" section (or "Discussion of Results") should respond to the statement listed as the "Purpose" or "Objective". Was the purpose successfully achieved? What was the final answer? Were there any problems in achieving the result?

If the objective was to determine the chemical oxygen demand of a water sample, the result should be stated as:

Results: The chemical oxygen demand (COD) of water sample #G-45 is 12 mg/L as determined by the standard dichromate method. Because the value was below the recommended range of accuracy, the analysis was repeated with a more dilute solution of the dichromate. This analysis resulted in a value of 14 mg/L that is considered to be more reliable than the first determination.

At times, the discussion of results overlaps with the conclusions. Using the same experiment (different results) as an example:

Results: The chemical oxygen demand (COD) of water sample #G-45 is 1220 mg/L as determined by the standard dichromate method. The sample had to be diluted to be in the reliable range of the procedure. After diluting, the sample remained turbid and this turbidity may have contributed to an erroneously high result.

Conclusion(s)

The Conclusions section should give an interpretation of the results. In many circumstances, making recommendations or suggestions regarding future work on the subject would also be appropriate. For the chemical oxygen demand determinations, the conclusions might be

> Conclusion: The low chemical oxygen demand (COD) of 14 mg/ L for sample #G-45 indicates that the water supply contains a very low amount of pollutants. Until results prove differently, all further testing of samples from this source should use the modified dilute dichromate method. The COD analysis does not discriminate between biologically degradable substances and nondegradable substances. I recommend that the sample be submitted for a BOD (biochemical oxygen demand) analysis and for coliform determination to resolve this matter.

or

> Conclusions: Sample #G-45 has a very high chemical oxygen demand (COD) of 1220 mg/L, a result indicating that it represents a severely polluted water source. Because the high turbidity of the sample may have contributed to errors in the determination, it is recommended that a comparison be made of the COD for a filtered and unfiltered sample. An odor of sulfide was apparent when acid was added to the sample during the COD determination; confirmation of S^{2-} would be valuable in characterizing the source of the pollution.

Reference

1. Kanare, H. M. *Writing the Laboratory Notebook*; American Chemical Society: Washington, DC, 1985.

Exercise

1. Using a previously written laboratory report and any additional guidelines issued by your instructor, note how you would rewrite it to improve the presentation of the experimental information.

| Chapter 13 |

Industrial and Business Reports

Almost every employed chemist will need to supply or prepare reports, usually serving a wide variety of needs and in response to a diverse number of requests. The documents may be simple memos providing desired information (*see* Chapter 18); letters to a colleague, customer, or client; or long, formal reports for use within or outside the organization. The reports could be proposals for funds, plans for research projects, recommendations about new product lines or policies (*see* Chapter 15), or publications of research results (*see* Chapter 14).

This chapter discusses reports that a professional scientist would be likely to encounter in his or her everyday activities within an organization, industry, or business. Before considering the components and formats of the report, let us determine the audiences or recipients of the report.

Audience

A report being prepared for publication in a professional journal is actually the easiest to write, because it will be read by peers and colleagues with equivalent (or nearly equivalent) backgrounds. Authors need not avoid using specialized vocabulary or jargon. Detailed explanations of background material are generally omitted or are pro-

1451–4/88/0123$06.00/1 © 1988 American Chemical Society

vided as references, and the author can get directly to the point and describe what was attempted, how it was attempted, and what the results were. (More information about preparing journal reports will be given in Chapter 14.)

In the industrial or business world, the scientist is likely to be part of a team that has members with a wide variety of educational backgrounds. The supervisor might not be a scientist, and the president of the company or members of the Board of Directors may not be well versed in the technical aspects of the scientist's work. If the report is being prepared for a client, it is likely that the client will need a clear explanation or interpretation of the results. These situations place a special responsibility on the scientist to provide the necessary information in a technically correct manner and to use explanations that a reasonably intelligent nonscientist can comprehend.

The task is similar to a social event where scientists are being introduced to various new acquaintances. If one chemist asks another what he or she is working on, he or she might reply, "I am trying new methods for synthesizing fluorescent analogs of 3,17-β-estradiol." If the new acquaintance is an accountant and asks the same question, the response would be very different: "I'm a chemist who works with hormones and similar compounds."

Components of a Formal Report

A formal business or industrial report may have all of the components given in the following paragraphs or may have only some of them. The style of the report may vary within an organization depending upon its purpose and intended use. A good report incorporates the following items.

Letter of Transmittal

If the report is to be used solely within an organization, it is typically accompanied by a cover memo (*see* Chapter 18). A very formal report to the president of the company, to the Board of Directors, or to an outside client or agency would have a letter accompanying it. (*See also* Chapter 16.) The memo or letter should present the report to the reader in the same manner that one would introduce a speaker to an audience. The writer should state how and why the report originated and at whose request. The scope and subject matter of the report should be mentioned, including any limitations or exclusions.

The letter is a perfect opportunity for the author to highlight important recommendations and conclusions or to bring the reader's attention to specific areas of the report. Unexpected findings or particular trouble spots or delays should be explained. Personal opinions are permissible, but they should be directed only to the findings of the report, and it is normally best to minimize them. Personal opinions should be clearly identified to ensure that the reader does not accept the statements as facts. The letter should conclude with acknowledgments of any assistance with the report and should contain a statement about possible follow-up actions, if any.

Positioning of the letter with the report may vary. If the report is being mailed, very often the letter is placed in an envelope and attached to the outside of the mailing package. When hand-delivered, the letter is often clipped to the report on top of the cover or is placed as the first page inside the cover. If the report is bound, the letter of transmittal is sometimes bound within the report, preceding the title page. Regardless of placement, the letter should be the first item noticed by the recipient (other than, perhaps, the cover of the report). A sample letter to accompany a report on a study of water quality in the Great Lakes is shown in Figure 1.

Cover and Title Page

Every report should have a cover, which can occasionally double as the title page. The cover can be colored or on stiff paper; some covers bear the logo of the organization. Printed on the cover should be the following information: title, date, to whom the report is being submitted, author, and business address. A project number or identification code should be included, if possible. An example is given in Figure 2.

Abstract

(*See* Chapter 9 for information about preparing abstracts.) The abstract should be an informative abstract and should stand alone in representing the contents of the report. It should be assumed that certain recipients will read only the abstract. The language should be

Environmental Testing Corporation

Box 1155

Rochester, NY 14607

(716) 271-0000

March 27, 1987

James Q. Watson, Director
District Water Quality Management
444 South Union Street
Rochester, NY 14607

Dear Mr. Watson:

As we agreed in January 1986, the Environmental Testing
Corporation accepted the task of monitoring the water
quality at 16 sites on Lake Ontario and 14 locations on Lake
Erie. The results of the analyses, conducted from March 1,
1986, to February 28, 1987, are included in the attached
report.

As expected, many of the parameters varied cyclically
with the seasons (see Appendix 1), and all of the parameters
varied with the depth of sampling (see Appendix 6). Please
note that there was no sampling for the month of August at
the Sodus Point site on Lake Ontario because the National
Regatta was held at that location; you and I agreed in July
that results taken during that event would be unreliable for
the purpose of this study.

The cooperation of the Coast Guard was vital in obtaining these results, and I appreciate your assistance in that
matter. After you and your colleagues have had time to
consider its contents, I would like to meet with you for
discussions about future testing as recommended in the
conclusion of the report.

I will be in contact with you in the middle of April to
arrange a time and place for our next meeting.

Sincerely yours,

John W. Wester

John W. Wester
Chief Chemical Engineer

JWW:ec

Enclosure

Figure 1. Letter of transmittal for a report.

```
ANALYSIS OF WATER QUALITY PARAMETERS:
      LAKE ERIE AND LAKE ONTARIO,
   MARCH 1, 1986-FEBRUARY 28, 1987




                Submitted to

          James Q. Watson, Director
      District Water Quality Management
          444 South Union Street
            Rochester, NY  14607




                    By


             John W. Wester
          Chief Chemical Engineer
      Environmental Testing Corporation


              March 27, 1987
```

Figure 2. Cover page (or title page) for a report.

aimed at a general reader, who is assumed to be an intelligent layperson but not a scientist. All major points, findings, and recommendations or conclusions should be included. In some companies, a report is filed by using the abstract, and 6–12 key words may need to be indicated for that purpose. Key words should be listed alphabetically.

Executive Summary

The practice of including an abstract and an executive summary has gained popularity, especially among executives. Although it means additional work for the author, an executive summary raises the odds that the reader will be familiar with the report. Unfortunately, many administrators do not have the time or do not find the time to read an entire report, especially if the report is long or technical and not commensurate with their backgrounds.

The executive summary is much longer than an abstract; it could be two to three pages, although the shorter it is, the better. It should be a complete summary of the project, including

- rationale for doing the project
- nature and scope of the study
- techniques used
- difficulties or problems encountered
- results obtained
- conclusions and interpretation of results
- recommendations

For the reader who thoroughly studies the entire report, the executive summary provides a quick overview before reading the report and presents a fast refresher when looking at the report later.

Table of Contents

The table of contents should be a separate page and should include a guide to all information contained in the report. Pages preceding the table of contents should be numbered using lower-case Roman numerals (i.e., i, ii, iii, iv, etc.), and pages following the table of contents should be numbered consecutively using Arabic numbers.

List of Tables, Illustrations, and Figures

This list should follow the table of contents, if appropriate to the report. Occasionally all tables, illustrations, and figures will be placed in the appendix, especially if extensive; a listing of the figures and other materials would then be included in the table of contents as titled appendixes (e.g., Appendix I: Cyclic Variation of Water Parameters).

Body of Text

The text should contain an introduction, the body of the material, and a conclusion. Headings and subheadings may be used as in an outline (*see* Chapter 5); this method is particularly effective in highlighting specific pieces of information. If headings and subheadings are used, they should be referred to in the table of contents.

The body of the text should include an introduction to the problem, review of any previously published studies related to the problem, explanation of methodologies or plan of attack, discussion of unexpected results or problems, presentation of results, interpretation of results (*see* information about this topic later in the chapter), and conclusions and recommendations. The last three items in this list are often set apart from the main body of the text in a section labeled "Summary", "Conclusions", or a similarly appropriate title.

The body of the text may contain data represented in tables, figures, graphs, or charts, and the text should refer specifically to their contents. If the data are extensive or if inserting them into the text interferes with the flow of the narrative, they can be placed in an appendix. Very often the supportive technical information is placed at the end of the report, especially if the primary audience is not well-versed in the topic.

Bibliography or Endnotes

If the report contains references to previously published material or if the information in that material would be useful to a reader who wanted to learn more about the subject, the report should have a bibliography, endnotes, or both (*see* Chapter 7).

Appendix

A report very often has one or more appendixes. One appendix could be a glossary that defines unfamiliar terms for the educated reader who may not have a background in the specific area covered. It might contain words that have special meaning in the context of the report, although these words are generally explained in the text. For example, "insoluble" may have different meanings to different people. The author might note: " 'Insoluble' in this report means solubilities of less than 0.0001 g of solute per 100 mL of water at 25 °C." Words in a glossary should be in alphabetical order; when they occur in the text, they should be highlighted by an asterisk or some other notation to indicate that the reader can find their definitions at the back of the report.

If the report is intended for nontechnical readers, the appendix is the best place to include the specific details of the study. They are present for reference, but will not obscure the reader's understanding of the text itself by being inserted as "barriers" throughout. Other materials that might warrant placement in an appendix include

- formulas or sample calculations
- extended details of experimental procedures
- samples of letters or exams
- samples of questionnaires or surveys used
- photographs or diagrams that are supportive of the report but do not require insertion into the text
- long lists of tabulated data
- resumes of authors or contributors to the report
- very long quotations, such as citations of court decisions, Environmental Protection Agency (EPA) guidelines, or Occupational Safety and Health Administration (OSHA) requirements

Index

A long report should contain an index so that the reader can quickly find specific information. The index should be arranged alphabetically. If the report is less than 20 pages and has a detailed table of contents with informative headings and subheadings, an index should be superfluous; in longer reports, it would be an asset.

Distribution

At some point in the report, it should be clearly indicated to whom it was given. If a memo is used to accompany the report, often a list of recipients is stated on it. If the list is not extensive, it could be provided at the end of the letter of transmittal by stating

Copies to: M. Jones
              G. Smith

or

cc: M. Jones
     G. Smith

or

Distribution: M. Jones
                     G. Smith

If the report is distributed to a large number of people, a separate page can be inserted into the report, listing each person and his or her organization or unit. This page could be placed at the end of the report or in the beginning before the title page.

Format

The format of a report is nearly as important as its content. If the format does not look attractive and neat, it will not receive the same treatment as one that is well organized and professional in appearance. Each organization has different guidelines regarding formats, but some rules are (or should be) universal.

✔ There should be no erasures, smudges, or type-overs.

✔ There should be no typographical errors; the report should be carefully proofread regardless of deadline.

✔ The type must be legible. The typewriter ribbon should be new, and the type should be letter quality in appearance.

✔ Margins should be uniform throughout the report with the possible exception of diagrams, tables, figures, and photographs; whenever possible, an effort should be made to fit them into the acceptable space. Top and bottom margins should leave at least 1 inch (2.5 cm). Normally, the left margin should be wider than the right to permit binding without obstructing the text. A suggested width is 1.5 inches (3.5–4.0 cm) for the left and 1 inch (2.5 cm) for the right. In addition to appearance, the margins provide the reader with an opportunity to make notes or to jot questions while using the report.

✔ The pages should all be numbered consecutively. The numbers can be at the top (centered or to the right) or at the bottom (centered).

✔ Headings should be used throughout the report and should follow the same guidelines as those used in outlines (*see* Chapter 5).

Data Interpretation

Although the audience might be well educated about the report's topic, the author has an obligation to provide an objective interpretation of the data presented. First, this practice may prevent someone from overlooking a point that the author believes is completely obvious. Second, it will help to prevent someone from misinterpreting the results. And third, the reader can agree with or dispute the author's viewpoint. An author's interpretation provides common ground for future discussion and action.

Many opportunities exist for providing an assessment of the data. Some examples are given in the following paragraphs:

- A comparison of different brands of equipment to determine which one provides the best service at the least cost should contain a conclusion by the author. After describing the features of each brand and the costs for accessories, installation, or training, the author should conclude how well each item meets the anticipated need of the organization and what drawbacks and benefits each would have. A personal opinion (based on the data) can be offered: "I suggest that we purchase Brand Q. Its many advantages, particularly that no special training is necessary for its use, completely outweigh the lower price of Brand R." This assessment is certainly preferable to simply comparing the items and leaving the conclusion to the reader.

- A report to a client definitely needs to include an evaluation of the results. If a water analysis was performed at the request of someone digging a new well, simply listing the numerical parameters that resulted from the laboratory work is not sufficient. For example, most people would not know what COD = 250 mg/L and BOD = 50 mg/L mean. In addition to listing the results, the author should include an explanatory paragraph, such as:

  The biochemical oxygen demand (BOD) is higher than is acceptable for drinking water and indicates pollution of the water source. Because this pollution could indicate contamination from sewage, I recommend that further testing be done on the biological activity (coliform). The high chemical oxygen demand (COD) is a matter for greater concern; the large difference between the COD and BOD shows the presence of a high percentage of pollutants that bacteria do not attack. This finding often indicates pollution from an industrial source. The water should not be used for drinking, and further testing for heavy metals, toxic chemicals, pesticides, and solvents should be conducted before the water is used for any purpose.

In a formal report that includes an executive summary, the interpretation of results should be incorporated in that section and in the main body of the text. This step ensures that the author's conclusions will be clearly understood by the audiences who receive and review the report.

| Chapter 14 |

Journal Publications

The speed at which scientific discoveries are being made and the potential importance of the results make it extremely necessary that the information be published for the scientific community to use, discuss, debate, substantiate, amend, or refute. As an example, the current public health problem of AIDS has resulted in publication of new results almost on a weekly basis: epidemiological statistics and interpretations, preliminary results of tests with new drugs, attempts to identify the causative agent(s) and concurrent trials of various vaccines, searches for animals that respond similarly to humans to both the disease and curative measures, new chemicals and preventive measures that may stem the spread of the disease, and more. Results of others' work must be disseminated so that various groups do not waste their time "reinventing the wheel".

Not all research results are viewed as being immediately useful like those just mentioned, but all research results contribute to the growing data base upon which all science derives its being. In addition, one reader often looks at someone else's findings and imagines ingenious uses of the results—uses that never occurred to the original researchers. If research results had not been published, many of the developments and conveniences that have led to our modern society might not have transpired.

If publishing is so important, how do you go about it? First, you need to have something to publish, and most scientific journals require that it be original work. Preliminary to doing research (in the laboratory or elsewhere), do a literature search in your area of interest for what has and has not been attempted. (See Chapter 6 for more details on this subject.) After you have clarified your goal, using the published information as a guide, devise a modus operandi as to how your problem should be attacked. If you find a solution to your problem and can confirm that it is indeed verifiable research, consider publishing the results. Most graduate students and researchers in the sciences can affirm that it can take several years of work (and usually some luck) before you have publishable results.

Steps To Publishing a Scientific Article in a Journal

Decide When To Publish

When should material be submitted for publication? As soon as possible after conclusive results have been obtained. Preliminary results are accepted by some journals, usually in the form of a "Communication to the Editor", but this format allows only for rapid publication of results of unusual urgency or significance. Normally, journals prefer to have the project (or one facet of a large project) completed before accepting manuscripts for publication.

Decide Where To Publish

A multitude of journals publish chemically related articles. If all of the sciences and engineering technologies are considered, the number of choices appears to be staggering. The choices narrow rather rapidly if the prospective author considers the audience who would be interested in the results. Obviously, organic chemists would not consider *Inorganic Chemistry* or *The Journal of Physical Chemistry* typical avenues for pursuing publication. Another guideline is to look at articles published by the selected journal: Does yours fit the general type of research being reported? In most journals, the editorial policy toward accepting submitted articles is stated in a "Notice to Authors of Papers" or similar set of guidelines, usually appearing in the first or last issue of each year or volume. Two examples follow.

> Inorganic Chemistry *publishes fundamental studies, both experimental and theoretical, in all phases of inorganic chemistry including the allied fields of organometallic chemistry, catalysis, and solid-state and bioinorganic chemistry. Emphasis is placed upon the synthesis and*

*properties of significant new compounds and structural, thermody-
namic, kinetic, and spectroscopic studies related to structure, bonding,
and reaction dynamics. A contribution must be original and must not
have appeared elsewhere.* (1)

The Journal of Organic Chemistry *invites original contributions
on fundamental researches in all branches of the theory and practice
of organic chemistry. It is not possible to publish all of the work
submitted to this journal, and, in the selection by the editors of manuscripts for publication, emphasis is placed on the quality and originality
of the work.*

*Papers in which the primary interest lies in the implications of
new compounds for medicinal, polymer, agricultural, or analytical
chemistry are generally considered to be published most appropriately
in specialized journals, together with information on evaluation with
respect to the original reason for synthesis.* (2)

Decide What Type of Publication To Submit

Journals accept several types of submitted material: articles, notes,
communications, reviews, correspondence, and other specialized publications that are specified in the "Notice to Authors". The most
typical include

Articles. "Full length articles should describe significant, complete
studies. An informative abstract not exceeding 300 words precedes
each Article. The title and abstract for an Article provide most of
the information upon which indexing is based." (1) "Articles should
be comprehensive and critical accounts of work in a given area." (2)

Notes. "Notes are significantly shorter than Articles; they should not
exceed 550 words. Extensive speculative interpretation in a Note is
inappropriate. Notes are not preceded by abstracts; however, an abstract should accompany the manuscript." (1) "Notes should be concise accounts of work of a limited scope. The standards of quality for
notes are the same as those for articles. Improved procedures of wide
applicability or interest, or accounts of novel observations or of compounds of special interest, often constitute useful notes. Notes should
not be used to report inconclusive or routine results or small fragments
of a larger body of work but, rather, work of a terminal nature." (2)

Not all journals accept notes; those that do are firm about the
rule that they represent completed projects. For this reason, notes
normally constitute only a small amount of the published material.

Communications. "Communications to the Editor are restricted to (usually preliminary) reports of unusual urgency, significance, and interest in all areas of inorganic chemistry and should be limited to 1500 words or the equivalent. A statement from the authors as to why their manuscript meets these criteria will be helpful to the editors in making their decision. Since no abstract is published, it is desirable that the principal conclusions be stated in the opening sentences." (1)

The word limit for communications varies from one journal to another (e.g., *The Journal of Organic Chemistry* has a limit of 1000 words and *Inorganic Chemistry* permits 1500 words). Although an abstract is not published with the communication, usually an abstract is requested and is published in *Chemical Abstracts.*

Reviews. Review articles are not accepted by some journals because they typically only include synopses of previously published results. Journals that do publish reviews normally solicit reviews that are generally directed to a specific topic, technique, or theme. In most situations, if an author wishes to write an unsolicited review, he or she would be wise to propose the idea to the journal editor for preliminary approval. An extensive bibliography and critical annotations are usually expected. Journals that accept review articles include *Chemical Reviews* and *Accounts of Chemical Research.*

Determine the Specific Format Required for Submission

The question of format, fortunately, is seldom left open; most journals are explicit about the manner in which the material should be presented, and a wise author will follow the published guidelines to the letter. Because of the many articles received by the journals, editors do not have the time to make extensive format changes and will frequently return the incorrectly presented article.

All manuscripts must be typed double-spaced on one side of good bond paper. Type *all* copy double-spaced, including footnotes and references, tables, captions, and the abstract. If the manuscript has a few errors, you may cross them out and type the correction above the error; if there are extensive errors on a page, completely retype the page. Use of correction fluid or correction tape is permissible; strike-overs or type-overs are not acceptable because they leave doubt as to which is the correct letter and sometimes cause the letter to be obscured. Normally, multiple copies (three, usually) are required at the time of submission because at least two copies are sent to peer reviewers.

Many publishers are encouraging authors to submit a diskette if the paper was prepared using a computer or word processor. Specific guidelines are given in the "Notice to Authors" or they can be obtained by contacting the editor. Usually, the diskette and one printout, double-spaced on continuous-feed paper, are acceptable. The cover letter should specify the brand and model of the system used and the name and edition of the word processing program.

Decide on the Organization of Material

By consulting the "Notice to Authors" and looking at several articles published by the targeted journal, the author should have little question about the sections that should subdivide the paper. Some journals, such as *The Journal of Organic Chemistry* (2), are very precise in defining the correct organization:

> An introductory paragraph or statement should be given, placing the work in the appropriate context and clearly stating the purpose and objectives of the research. The background discussion should be brief and restricted to pertinent material; extensive reviews of prior work should be avoided; and documentation of the literature should be selective rather than exhaustive, particularly if reviews can be cited.
>
> The discussion and experimental sections should be clearly distinguished, with a separate center heading for the latter; other center headings should be used sparingly. The presentation of experimental details in the discussion section, e.g., physical properties of compounds, should be kept to a minimum.
>
> All sections of the paper must be presented in as concise a manner as possible consistent with clarity of expression. In the Experimental Section, specific representative procedures should be given when possible, rather than repetitive individual descriptions. Standard techniques and procedures used throughout the work should be stated at the beginning of the Experimental Section. Tabulation of experimental results is encouraged when this leads to more effective presentation or more economical use of space. Spectral data should be included with other physical properties and analyses of compounds in the Experimental Section or in tables. Separate tabulations of spectral values should be used only when necessary for comparisons and discussion.

In general, the paper should have an introductory section with a statement of the problem, the significance of the problem, and its relationship to current research. The introduction should point the reader to existing pertinent knowledge about the subject, but usually should not have an extensive literature review, although some background can be incorporated by citing previously published review articles.

If the research involved laboratory work, an Experimental Section should be included. It should be written with sufficient details so that anyone could repeat the experiment and obtain the same results. If materials, procedures, or apparatus are standard, they do not require much description; frequently a reference to a procedure will suffice. The accuracy of the measurements should be mentioned, and any possible experimental hazard should be identified, normally with the word CAUTION, as shown, in bold capital letters, followed by a statement.

The conclusion of the paper is often partitioned into sections called "Results" and "Discussion"; sometimes the sections are combined. Material represented previously in figures or tables does not need to be repeated, but it is appropriate to state the results, their limitations, interpretions of their meaning, and comparisons with the results of others. Areas that require further examination or experiments that should be conducted may be mentioned as future recommendations.

Finally, in the acknowledgment section the author may wish to refer to the contributions of others (some publishers suggest not including typists, secretaries, or artists). Perhaps a colleague made a suggestion that changed the direction of the research; without his or her suggestion, your results would not have been obtained. If the colleague is not a coauthor, you might want to acknowledge her or him. Frequently, funded researchers state the grant number and granting agency in the acknowledgment section; an alternative is to place the information in a footnote. (Often the granting agency specifies the wording that should be used.) If a preliminary presentation of the published material was given at a professional meeting, mention it here or in a footnote.

Review Process

When the manuscript has been completed and typed according to the journal's specifications, it should be very carefully proofread. It helps to have a colleague assist with proofreading; he or she can also give you advice about the organization of the paper, missing details, or ambiguities.

If the author is satisfied with the result, then the required number of copies should be mailed (first class) to the publisher, accompanied by a letter of transmittal. (Some authors insure the mailing; all authors should keep a copy for themselves!) The letter should state your intention that the manuscript [name the title and author(s)] be considered for publication. If you feel that the editor might have some doubt about your choice of journal, you might want to explain why you selected the particular journal for your article. If the subject is one that the journal normally publishes or if you have previously published in that journal, you do not need to explain your choice at length.

The letter should clearly state the author to whom all communications should be sent. This information is very important when it is time to review the proofs. If a copyright waiver form is required for publication, it should be enclosed with the letter. A statement confirming that the manuscript has not been published previously and is not being considered for publication in other journals should be made. (A sample letter of transmittal is shown in Chapter 16.)

Upon receipt of the manuscript, the editor will screen it to make sure it is appropriate for the journal. If it is, the editor will send the manuscript to at least two reviewers who are experts in the subject. Although the criteria for review vary by journal, most judge the merit of the paper, its significance, the validity of the measurements and of the interpretation of the data, and other supporting information. The reviewers write comments about the submission and return them to the editor, who in turn sends copies to you (typically without revealing the reviewers' identities). The editor relies upon the reviewers' recommendations to publish (or not), although some circumstances permit the editor to make a decision contrary to the reviewers' suggestions. If the paper is rejected, reviewers or editors will often give helpful comments about improvements needed for resubmission.

If the paper is accepted, it is copy edited and sent to the typesetter. Typeset galleys are proofread by both the editor and the author, who needs to give final approval before the manuscript will be published. The proofreading process is vital to the integrity of the publication; the author should be very thorough in this effort. (*See* Chapter 11 for more specific information about proofreading.) Lengthy delays or extensive revisions by the author, however, may delay publication. Massive revisions at this point might result in the manuscript being sent back to reviewers, and the entire process would begin again. In addition, some publishers charge the author for the expense of resetting the type.

A form for reprints (if this is a paper) will be mailed with or shortly after the galley proofs. The author should order the desired amount and include a purchase order.

If the journal has page charges, authors will be invoiced for them, also with or shortly after galley proofs are sent. If errors are found in the published manuscript, a letter should be sent to the editor of the journal as soon as possible. Most journals publish letters with Errata in a timely fashion.

References

1. *Inorg. Chem.* **1986,** 25, 4A.
2. *J. Org. Chem.* **1986,** 51, 9A.

| Chapter 15 |

Grants and Proposals

A professional scientist is almost certain to encounter a time when he or she needs to write a proposal. You or the organization for which you work may want to apply to a governmental agency or to a foundation for funds to perform a special project, to obtain equipment that could not otherwise be purchased, or to sponsor research efforts in a specific area. A proposal need not be limited to these situations, however.

If you are working for an industrial organization, you might want to write a proposal outlining a new method for performing an existing activity or suggesting the best piece of equipment to be purchased to implement a new product line. Your proposal could be directed to changing a company policy; for example, you might want to illustrate how a four-day work week would be more advantageous to the firm than the existing five-day norm.

As a private citizen, you might submit a proposal to the municipal government supporting changing the street on which you live into one-way traffic. Perhaps you want to convince officials that extending public transportation to your neighborhood would be beneficial to the downtown district. Or maybe you would like to advocate the benefits of year-round public school education.

All of these situations require a written document in the form of a proposal, but the format and complexity would vary from one situation to another. Generally, the complexity will increase with the amount of funding needed to accomplish the proposal, although sometimes you need to be very specific and detailed when the project is relatively inexpensive.

1451–4/88/0143$06.00/1 © 1988 American Chemical Society

Types of Proposals

A solicited proposal, such as one for a National Science Foundation (NSF) grant or Ford Foundation grant, will be advertised or publicized in appropriate scientific journals or governmental publications. The announcement usually summarizes the nature of the desired proposals, deadline date for application, and an address where more information can be obtained. The larger granting institutions often have a pamphlet or booklet containing guidelines for submitting proposals. Typically, if you are applying to a large, well-established organization, you will complete standard forms that have rather detailed instructions on the information required.

For example, Figure 1 shows a recent Program Announcement and Guidelines (1) for the NSF "College Science Instrumentation Program". An unsolicited proposal would not have *all* of the prepared materials as indicated in this rather thorough treatment. Depending on the scope of the project, the unsolicited proposal should contain many of those components.

Proposals can be categorized as internal or external. An internal proposal would be used only within the writer's organization and would need less background information than a proposal that would be sent to another organization. An external proposal would need to provide details, such as history of the organization and qualifications of the proposal author, which might be unnecessary for in-house use.

Proposals can be grouped by their purpose or intent. A research proposal would seek approval of a prospective direction for experimentation. It could be an internal, solicited proposal; for example, a student may suggest an area of research designed to lead to a Master's or Ph.D. degree. The research proposal could be internal and unsolicited—as simple as a contribution to the company's Suggestion Box or as complex as trying to convince the organization to change its product preparation techniques. Many research proposals are external (solicited or unsolicited) and most often involve a request for outside funding for equipment, supplies, or salaries (or a combination of the three).

A sales proposal is most often external. You might be trying to convince clients to use your product or your services to accomplish a stated goal or to improve their present situation. This proposal could be a response to an advertised request for competitive bidding (solicited) or an effort to initiate business with a specific prospective client (unsolicited).

I. Program Description
   A. Purpose and Scope
   B. Eligibility Criteria and Limitations
      1. Eligible Institutions
      2. Eligible Fields
      3. Eligible Departments
      4. Eligible Activities
      5. Eligible Equipment
      6. Ineligible Items
      7. Eligible Project Size
   C. Requirements for Matching Funds

II. Preparation and Submission of Proposals
   A. General Information
   B. Proposal Preparation
   C. Proposal Format
      1. Cover Sheet
      2. Project Summary Form
      3. Detailed Budget
         a. Scientific Equipment
         b. Computing Equipment
         c. Equipment Fabrication
         d. Safety Equipment
         e. Maintenance Equipment
         f. Shipping Costs
      4. Narrative
         a. The Present Situation
         b. The Development Plan
         c. Equipment
         d. Faculty Expertise
      5. Appendices
      6. Proposal Submission

III. Proposal Evaluation and Award Selection

IV. Announcement and Administration of Awards

V. Check List

VI. Other Programs

VII. Forms

Figure 1. A recent Program Announcement and Guidelines for the NSF "College Science Instrumentation Program".

Another type of proposal is called a planning document. A planning document could also be either internal or external, solicited or unsolicited. For example, you may suggest that the organization computerize its production line. If approved, a planning proposal often leads to a sales or research proposal or a combination of all three to accomplish the desired goal.

Parts of a Proposal

A good proposal generally has the same elements as any good report. It must have an introduction, the body of the proposal, and a conclusion. An abstract or a summary are required for most proposals.

The author of a proposal needs to consider the audience and its background. If the primary readers are peers, the contents can be very technical and vocabulary specific to the scientific area can be used without extensive definition. If the background of a secondary audience is not well established, a glossary of technical terminology or other supplemental information can help the audience understand the proposal. Often the primary audience is not well informed; in this situation, the body of the proposal should be specific but less technical, and an appendix of the detailed technical descriptions should be provided for a secondary audience of experts.

Format of the proposal is important. Published guidelines (if they exist) must be followed to the letter, and all items requested must be included in the final copy (or a statement explaining why the materials are not present). As with all technical writing, the proposal must be in perfect, grammatical English and must not have any typographical errors, strike-overs, crossed-out letters, or white-out paint. The proposal should be prepared on white bond paper unless colored forms are provided or specified (e.g., NSF requires colored paper for appendixes so that they can be easily distinguished from the rest of the proposal), and fancy folders or binders or other "gimmicky" covers should be avoided. The final product should be attractive, neat, and somewhat conservative in appearance.

The title of the proposal is extremely important. It should state clearly what the proposal is about in as few words as possible. A longer title is preferred if a shorter version would be ambiguous. The following title for a proposal is short and neat but not very informative:

A Proposal to Study AIDS

The reader would have no idea about the focus or target of the suggested study because the topic encompasses such a large area. A more suitable title might be

A Proposal to Gather Statistical and Epidemiological Data
on AIDS Patients in Rochester, NY

This much longer title gives the reader enough information to decide if he or she wants to read further.

An external proposal should have a cover sheet that lists the title of the proposal, name(s) and affiliation(s) of the author(s), the date, and where or to whom the proposal is being sent. Frequently, the signatures of the principal investigator, head of the organization, or both are requested on the cover sheet (or elsewhere in the proposal). A cover letter of transmittal (*see* Chapter 16) normally accompanies the external proposal. An internal proposal may require a cover sheet, depending upon the complexity of the proposal and policies of the organization; a simple internal proposal might be presented adequately in memo form (*see* Chapter 18).

The summary or abstract should be as brief as possible (*see* Chapter 9) but complete in its description of the project. The audience needs to be considered especially in this respect. Many foundations use the summary to determine appropriate peer reviewers; therefore, the author can be technical in indicating the problem being addressed, general methods and needs, and the significance of anticipated results. Because governmental grants are open to public inspection, agencies such as NSF warn (1) that summaries "*should be written with considerable care so that a scientifically literate layperson can understand the use of Federal funds in support of the project.*" [Italics as published.]

A detailed budget needs to be included if any funds are being requested. Usually the budget is heavily referenced, with several explanatory notes for the various items in the budget being presented. Equipment costs are usually documented by a catalog price or manufacturer's quote. Salaries are estimated according to hourly, weekly, or monthly rates with a time estimate included; if benefits are to be incorporated, they should be specified. Travel expenses (if allowed) should be explained; number of trips, destination, and purpose should be listed. The budget should be as realistic as possible. Consider providing possible reasons why a cost overrun could occur and what contingencies are available for dealing with a cost overrun. Include the explanation in the budget section or a separate section.

The body of the proposal is the most important section because it is the primary basis on which the reader will decide to accept or reject the proposal. The introduction should contain answers to the following questions, as appropriate:

- What is the present situation?
- What problem do you want to solve?
- Why are you trying to solve the problem?
- What will the outcome of the project provide that will be beneficial (and to whom)?
- If the project is research oriented, what has been done by you (and others) related to your proposal (e.g., a literature review)?
- What are your special qualifications for attempting the project?

The body should be detailed enough to provide complete information, but should not be wordy. The people who will review the request will have to read many proposals and are likely to be impatient with an overly long submission. (Some solicited proposals have page length requirements described in the program announcement or guidelines.) Most of the following questions should be answered:

- What is the anticipated length of the project?
- Is there a timetable (time analysis) for various steps toward completion of the project?
- Is there a standard protocol or procedure for accomplishing the stated goal?
- What is required to achieve the goal? personnel? equipment? new methodologies? special materials? special facilities?
- Have alternatives been considered (or rejected)?
- What will be used to measure the success (or failure) of the project?

The conclusion or summary should give an overview of the project and an emphatic statement explaining why you or your company should be chosen for this award. It should be brief but comprehensive.

Most supporting material should be in appendixes rather than in the body of the narrative. The material contained in the appendixes can vary greatly depending upon the requests of the granting agency,

the nature of the grant, and the financial amount of the grant. Typically, items found in the appendixes are the following:

- a detailed budget with an explanation of the various items (if
  the explanation is not incorporated into the body of the proposal)

- a resume (or vita) for the principal investigator and others
  who would be involved in the project (*see* Chapter 17)

- a list of equipment available to perform the project (Equipment
  needed, but not available, would generally be listed in the
  budget section.)

- a list of successfully completed projects or former grants

- a statement of current support (especially if the request is for
  a continuation or extension of an existing grant)

Reference

1. *Catalog of Federal Domestic Assistance Number 47.064 (College Science
   Instrumentation Program)*; National Science Foundation: Washington,
   DC, 1986; NSF 86–23, OMB 3145–0058.

Bibliography

Mertens, T. R. "Reflections on Writing and Reviewing Grant Proposals";
J. Coll. Sci. Teaching **1987,** XV(4), 267–269.

Exercises

1. Write a proposal to the department head (chair) or curriculum committee
   regarding one of the following:
   (a) changing the content of an existing course
   (b) requesting the addition of a new course
   (c) eliminating the requirement of an existing course
   (d) changing one of the departmental policies
   (e) requesting money for a student organization for a special project

2. If you have (or have had) a part-time job, write a proposal to your

supervisor addressing one of the following (include cost estimates or savings, if appropriate):

(a) making customer service more effective
(b) a new advertising campaign
(c) a change in company policy
(d) an addition to the benefits package
(e) a reorganization of your work unit

| Chapter 16 |

Business Correspondence

Although telephones, electronic mail, and computers have greatly reduced the number of business letters that most professionals have to write, business letters are still the preferred means of communicating when a permanent record is desired. There will be many instances in a person's scientific career when he or she will need to write correctly styled professional letters. The level and nature of the position will determine how often this responsibility occurs.

Everyone needs to write at least one professional letter—the letter that accompanies the résumé when applying for employment. A scientist submitting a journal article for publication needs to write a letter extolling the merits of the article to the editor of the journal. A technician may need to write a letter accompanying an analysis of a sample to explain the results to a client or to the supervisor. Because in each situation, an enclosure accompanies the letter, it is generally called a cover letter (or less frequently, covering letter).

There are many other reasons for writing business letters. For example, to invite a noted scientist to speak, a letter is necessary to extend the invitation, another to confirm arrangements (or to express disappointment if the invitation is not accepted), and another to express gratitude after the visit. In addition, scientists frequently write to one another requesting and exchanging information about mutual projects or similar projects.

Someone in an administrative or supervisory position may need to write a variety of letters, all requiring exact and correct phrasing. For example, a contractual letter will be written to offer a position

1451–4/88/0151$06.00/1 © 1988 American Chemical Society

to a successful applicant or to authorize the purchase of a major piece of equipment. Sometimes an administrator or supervisor will need to write letters involving recommendations, complaints, congratulations, apologies, and condolences.

Additionally, a letter extends a more personal touch to a communication and could be used to confirm a phone conversation or electronic mail decision.

Regardless of the purpose of the business letter, all have specific requirements in common. A business letter is always typewritten, normally on letterhead paper from the company or institution. It should contain no typographical errors, and grammar should be perfect. However, you can be less formal with people you know well or with whom you communicate often.

Parts of a Business Letter

Figure 1 shows the layout of a typical business letter on letterhead paper. The stationery of the company or institution has the return address printed on it; if the writer's address is not engraved on the letterhead, it should be the first item on the page. The date follows the writer's address.

After a double-spaced line, the name, title, and mailing address (with ZIP code and building or room numbers, as appropriate) of the person being written to are listed with no punctuation at the ends of the lines. Abbreviations (St., Ave., Dr.) should be avoided in the inside address, although official Postal Service two-letter designations for states may be used (without periods).

The salutation appears next, separated from the mailing address and the body of the letter by double spaces. A colon is always used after the salutation. Address a person by his or her title in the salutation (i.e., Dr., Prof., Mr., Ms., Mrs., Dean) unless you and the addressee are on a first-name basis. If you are not sure how to address the person, using a person's title is always correct. Using the person's name is always preferred, but sometimes, such as when writing a letter of complaint to a company, it is not easy to obtain the person's name. In these situations, you may have to resort to writing impersonal salutations such as "Dear Sir or Madam" or "Attention Consumer Affairs Office".

The body of the letter should be as brief as possible without omitting any necessary information. If a letter is too short, the reader

DUDLEY'S ENVIRONMENTAL LABS
1205 NE Old Street, Cincinnati, OH 45202
Phone: (513) 777-0000

Date

Addressee's name, title
Business organization
Street Address or P.O. Box
City, State   ZIP code

Dear (Dr., Mr., Mrs., Ms., Prof., Dean, etc.) Last name:

The first paragraph is usually very short and has one or two
sentences to introduce the writer, if necessary, and the
purpose of the letter.

The body of the letter should describe the purpose in
complete, but brief, detail. Under most circumstances, the
body should contain one to three paragraphs. Paragraphs
should be separated by double spaces, but typing within
paragraphs should be single-spaced. Each paragraph should
contain several sentences.

There should be a closing statement, regarding expectations
or expressing gratitude, depending upon the purpose of the
communication.

Sincerely yours,

Your signature

Your name, degree (if appropriate)
Job title (if appropriate)

Typist's initials (if you are not the typist)

Enclosures or Attachments (if appropriate)

Figure 1. Typical business letter (nonindented).

might think the writer is being rude or curt, or is not concerned with the matter being discussed. The letter should be no longer than one page, although some circumstances require a longer letter (such as a response to a request for information). Typing should be single-spaced with a clear line of space between paragraphs.

The first paragraph should extend cordialities that may be desired (e.g., "It was a pleasure talking on the phone with you yesterday", "Thank you for your letter of September 8, 1986", or "I have read your most recent article and it was of great interest to me"). Usually, the first paragraph is short and should give the purpose for writing.

The conclusion should be a short paragraph, sometimes consisting of one sentence. Statements such as "If you need more information, please don't hesitate to write or call", "Thank you again for the inspiring lecture you gave to our group regarding your research", or "I am anticipating our next meeting at the ACS convention in Chicago" should suffice.

The closing should be one of the standards, such as "Sincerely", "Sincerely yours", or "Yours truly", followed by a comma. Your name followed by your title (if any) should be typed below the closing, leaving enough space for your signature to appear between the closing and your name. Your title never precedes your name, and is not included in your signature, although many people in the medical profession include "MD" in their signatures, perhaps from the habit of signing many papers daily that require it.

Women without titles such as Dr., Prof., or Dean may wish to indicate how they prefer being addressed by including Miss, Ms., or Mrs. in parentheses. For example,

Sincerely yours,

Joy Wu

(Ms.) Joy Wu, LPN
Coordinator

A postscript may be added after the signature. Postscripts are used frequently in advertising letters to emphasize a special offer, or to highlight a particular point in a regular business letter. Under normal conditions, it is best to use postscripts sparingly by including all necessary information within the body of the letter.

If the letter has been typed by someone other than the correspondent, the typist's initials should appear at the left margin after the closing. The format varies slightly from one institution to another, but the following are typical:

BEC:nh (initials of writer:initials of typist)

BEC/nh (initials of writer/initials of typist)

nh (initials of typist only)

Enclosures accompanying the letter should be indicated below the typist's initials at the left margin. ("Attachment" is used if the material is clipped or stapled to the document, a practice more frequent with memos. *See* Chapter 18.) Indicating enclosures or attachments alerts the reader to look for them and to notify the writer if they have been omitted. When multiple copies are enclosed, specify the number or identify them by name. Examples are as follows:

Enclosure

Encl.
Enclosures (2)

Enclosures 2

Encl. 2
Enclosures: College Catalog
           Application Blank
           Housing Information

Formats for Business Letters

Figures 1 and 2 show typical formats for business letters. The block or nonindented style illustrated in Figure 1 is gaining in popularity. The standard format with indented paragraphs and with the date and closing starting at about center page is shown in Figure 2.

DUDLEY'S ENVIRONMENTAL LABS
1205 NE Old Street, Cincinnati, OH 45202
Phone: (513) 777-0000

Date

Addressee's name, title
Business name
Street Address or P.O. Box
City, State   ZIP code

Dear (Dr., Mr., Mrs., Ms., Miss, Dean, etc.) Last name:

The first paragraph should give the purpose of the
letter and, if necessary, an explanation of who is writing.

The body of the letter should consist of one to three
paragraphs. The content should be a complete, but brief,
description of the purpose or intent. All paragraphs should
be separated by double spacing.

Typing within the paragraphs should be single-spaced.
Each paragraph should have several sentences. Try not to
write paragraphs that are too long, however.

The closing statement usually consists of one sentence
(maybe two) expressing expectations or otherwise indicating
how the intent of the letter will be carried out.

Sincerely,

Your signature

Your name, degree
Job title

Typist's initials

Enclosures or Attachments (as appropriate)

Figure 2. Typical business letter (indented).

Variations on both styles are acceptable, and most industries and businesses have adapted a format that is used by everyone representing them. Some organizations use the block form and center the date at the top of the page underneath their printed letterhead. Other companies use the block style with the exception of an indented closing and signature. The examples in Figures 1 and 2 represent the two extremes: completely block and fully indented.

General Considerations and Grammatical Style

The tone and language of the letter should be tailored to its purpose and to the background of the addressee or reader. Use a conversational tone without resorting to slang. Imagine how you would converse with the person on the phone and write the message in a similar style and manner. If appropriate, mention something personal so that the letter does not appear to be extremely formal.

Pronouns in letters seem to cause special problems. For some reason, many people fiercely avoid using "I" in a business letter. There is absolutely no prohibition on using "I"; it lends a more personal flavor to the letter. "I want to congratulate you on the publication of your latest book" is much more pleasant than "You are to be congratulated on the occasion of your book's publication."

Caution should be observed in the use of "we". Three problems are associated with this pronoun. First, "we" can often be ambiguous—who is "we"? "We" could be the company, the division, the people in the office, or the writer's family. Second, if "we" is meant to be the company, you might be committing your institution to something that they may not wish to endorse. And third, most letters are written and signed by one person (you), and by using "we" in the text of the letter, the text may sound pretentious (as in the Queen's "we" or the editorial "we"). Worse, it may be grammatically incorrect.

The pronoun "you" can be used too frequently. For example, we have all probably received communications that state: "As you already know. . . ." If you already know, why should you read the rest of the letter? Consider the phrase "You will, no doubt, wish to. . . ." or "You will certainly not want to miss this opportunity to contribute to. . . ." These statements come across as demands, and the reader may stop reading the letter and throw it away!

Occasions for Writing Business Letters

Cover Letter for a Job Application

If you are responding to an advertisement for a specific job, your letter should address the items mentioned in the job description and how you qualify for the position. The cover letter should always be accompanied by a résumé (see Chapter 17), and it should highlight or expand on the qualifications listed in the résumé.

Unless the advertisement is a "blind ad", (i.e., the ad does not name the hiring company or institution) your letter should be addressed to the person named. If a person's name is not mentioned, call the company to get the name of the person who will review the applications. It is always best to identify the person in the salutation of the letter. Otherwise, you can only resort to the very impersonal forms, such as "Dear Sir or Madam" or "To Whom It May Concern", or the sexist forms "Dear Sir" or "Gentlemen". Another alternative would be to use "Attention: Personnel Director" instead of a salutation. Clearly, having an actual name to use is a much better practice.

If you are not applying in response to an advertised position (as is the situation for many people approaching graduation), it is still a good idea to learn the name of the appropriate person in the company or institution. A letter simply addressed "Personnel Director" or "Head, Science Department" runs a much greater risk of being discarded or unopened than one addressed to a specific person.

After describing why you believe the company or institution should consider you a qualified applicant for the position, end the letter with a statement that you are available to be interviewed. Write how you can be contacted. Mention that your résumé is enclosed and that you would be happy to provide any other information. If the position is not in the city in which you live, mention possible times that you plan to be (or could be) in the employer's neighborhood.

As with the résumé, the cover letter must be free of errors—no typographical or grammatical mistakes. The letter (and envelope) must be typed. The paper must be of good quality (bond, preferably) and should be white or perhaps off-white.

A typical cover letter for a job application is shown in Figure 3. Your letter should reflect you and, therefore, the example given is simply a guideline. Your letter can (and should) show your individuality and imagination. Remember that you are trying to convince the employer that you are worthy of consideration for the position.

```
                                        Box 000
                                        Main Post Office
                                        Rochester, NY  14623
                                        September 8, 1987


Mr. James D. Houseman
Division Manager
Syracuse Chemical Company
South Salina Street
Syracuse, NY    13201

Dear Mr. Houseman:

     The position of analytical chemist, which was advertised in
Sunday's Rochester Democrat/Chronicle, interests me very much.
My course work as a Chemistry major at Rochester Institute of
Technology (RIT) and my work experience at IBM have provided me
with the necessary qualifications for the opening.

     I expect to receive my baccalaureate degree (certified by
the American Chemical Society) in December 1987.  Analytical
chemistry has been stressed throughout the curriculum at RIT, and
I have taken the courses Introduction to Chemical Analysis,
Instrumental Analysis, Separations Techniques, and Advanced
Instrumental Analysis.  My resume (enclosed) shows other relevant
course work.

     For the past three summers, I worked as a chemical tech-
nician at IBM, and my responsibilities included analytical work
in the quality control area.  Last summer, I learned the proce-
dures for analyzing environmental samples. I found that to be an
area in which I would like to get more experience.  Duane Meyers,
supervisor of quality control, rated my performance as
"excellent" for all three summer positions and was complimentary
about my initiative and willingness to learn new techniques.

     I would appreciate an opportunity to discuss the available
position with you.  I can be contacted by phone at (716)
475-0000, mornings, or (716) 271-0000, afternoons.

                                   Sincerely,

                                   *Agnes L. Barry*

                                   Agnes L. Barry

Enclosure
```

Figure 3. Sample cover letter.

Follow-Up Letter

If you are fortunate enough to be granted an interview, you should immediately (no later than the day after the interview) write a letter of appreciation to the person who hosted you. The letter should conform to business letter style and should provide specific details about the visit. If you are still interested in the position, mention that you are (do so with enthusiasm). Figure 4 shows a simple follow-up letter that conveys continued interest in the position. (If you are not interested after the interview, say gracefully that your name should be removed from consideration; you do not need to give a reason.) [See the section "Letter Declining an Offer".]

Letter Accepting an Employment Offer

Congratulations! You have been offered a job and you wish to accept it. Immediately, if possible, write a letter of acceptance. Your letter should repeat the factors regarding your employment: your salary, starting date, and any other important details. If the offer of employment contained any questions, you should naturally respond to those as well. The letter does not need to be long or emotional. You simply want to say that you accept the offer and under what conditions you are accepting it.

Occasionally the situation is more complex. What if you are not completely satisfied with the offer? Perhaps the salary cited is lower than you expected, the benefits are not sufficient, or the job title or level is different than what you anticipated. In any of these situations, you should state clearly that you want to discuss the issue in more depth before you formally commit yourself. You may be jeopardizing your opportunity, but any part of the offer with which you are ill at ease should be resolved before you accept it. If agreement cannot be reached, perhaps there was not as good a match between you and the prospective employer as you both thought. Figure 5 gives an example of a typical acceptance letter.

Letter Declining an Offer

Sometimes an offer of employment is not suitable—or, if you are fortunate, you might have more than one offer! Your decision should be made as soon as possible, and you should respond to the employer in a prompt, courteous manner. If you are comfortable giving a reason for refusing the offer, do so; if the reason has anything negative about the company, omit it.

The best refusal is usually short and polite. Acknowledge how much you appreciate the offer and how sorry you are that you have to decline. Try to leave the company with a good feeling about their choice (you). A sample letter is shown in Figure 6.

Box 000
Main Post Office
Rochester, NY 14623
October 27, 1987

Mr. James D. Houseman
Division Manager
Syracuse Chemical Company
South Salina Street
Syracuse, NY 13201

Dear Mr. Houseman:

Thank you for the opportunity to meet with you and Ms. Handy yesterday. I appreciate the time that you both spent with me, and I was very impressed with your well-equipped laboratories.

Everyone I met was most cordial in answering my questions, and the atmosphere at your company has increased my interest in the analytical chemistry position. The project that your division is working on would be a fascinating challenge for me. I am confident that I have the skills to contribute to your efforts.

If you need any more information regarding my experience and background, please contact me at any time. I anticipate hearing from you soon.

Sincerely,

Agnes L. Barry

Agnes L. Barry

Figure 4. Sample follow-up letter.

```
                                        Box 000
                                        Main Post Office
                                        Rochester, NY    14623
                                        December 1, 1987


Mr. James D. Houseman
Division Manager
Syracuse Chemical Company
South Salina Street
Syracuse, NY    13201

Dear Mr. Houseman:

    I am very happy to accept your offer of employment as
an analytical chemist in your division.  I appreciate your
kind letter of November 25, 1987, extending this opportunity
to me.

    As specified in your letter, I will be able to start
working on January 2, 1988, at the annual salary of $25,800.
I understand that I should report to the Personnel Office at
8:00 a.m. to complete the necessary paper work.

    I will be moving to Syracuse at the end of the month
and will notify you of my new address and phone number when
arrangements are complete.  I am very enthusiastic about
joining the Syracuse Chemical Company on January 2.

                                    Sincerely,

                                    Agnes L. Barry

                                    Agnes L. Barry
```

Figure 5. Sample letter of acceptance.

Letter for Transmitting a Manuscript (or Abstract for a Talk)

If you are submitting a manuscript for publication or an abstract for
an oral presentation at a conference or convention, you need to do
so by letter. Unless the submission is an invited paper (i.e., requested
by the editor or conference organizer), indicate to the addressee why

your paper is important enough to be considered. In addition, state the reasons that you think the audience will be interested in the subject.

If the intended presentation is to be published, indicate that it has not been previously published. Many journals will also request that you submit a signed form that releases copyright to them. Usually this information is contained in the journal's "Notice to Authors", which is typically in the first or last publication of the calendar year.

If the paper has multiple authors, make it clear which person should be contacted for further communication. For a proposal for an oral presentation, the letter should contain an estimate of the time needed (or desired) to give the presentation and any special media requirements (e.g., slide projector, overhead projector, demonstration table, or chalkboard). Figure 7 shows a typical letter accompanying a manuscript and could be suitably adapted to a proposal for an oral presentation.

Contractual Letter

Often an administrator needs to write a letter that is equivalent to a contract. This letter may be an offer of a permanent position, a bid to purchase some major equipment, or an acceptance of the services of a consultant. A letter of this nature should be very formal and should give the conditions explicitly. If an actual contract form is not used, the letter should be read and approved by a colleague or supervisor (or preferably, a legal expert). Sometimes the letter should be set aside for a day or two so that any second thoughts can be exercised.

Letter of Complaint

A letter of complaint can be written for many reasons. For example, you may wish to write to the editor of the local newspaper, to the utility company, to a contractor, or to an employee (or boss). Letters of complaint can be very difficult to write. When you are angry or upset, you may not present your case logically, and the text can become unnecessarily wordy. The most effective letter of complaint states the problem clearly and unemotionally. A courteous, businesslike manner should be maintained even if the situation is extremely frustrating. All of the important circumstances should be

```
                                        Box 000
                                        Main Post Office
                                        Rochester, NY    14623
                                        December 1, 1987


Mr. James D. Houseman
Division Manager
Syracuse Chemical Company
South Salina Street
Syracuse, NY    13201

Dear Mr. Houseman:

     Thank you for your letter of November 25, 1987, offer-
ing me a position as analytical chemist with the Syracuse
Chemical Company.

     After much deliberation, I am sorry that I must decline
the offer.  Your company, facilities, equipment, and person-
nel all impressed me very much, but my career goals have
shifted recently and I have decided to continue my education
by pursuing the doctorate degree in Chemistry on a full-time
basis.

     I appreciate your confidence in me.  Thank you again
for the offer of employment.

                                    Sincerely,

                                    Agnes L. Barry

                                    Agnes L. Barry
```

Figure 6. Sample letter of refusal.

described, but the letter should be kept as short and to the point as possible. A lengthy, complaining letter is easily set aside by the recipient or, as occasionally happens, is sent to a superior, where it might remain on a desk for some time.

Letter of Apology

Although it is difficult for us to admit, we all make mistakes, and at times we have to apologize to others formally. A letter of apology

Department of Chemistry
Rochester Institute of Technology
Rochester, NY 14623-0887
September 9, 1987

Dr. Lester G. Paldy, Director
Center for Science, Mathematics,
    and Technology Education
State University of New York
    at Stony Brook
Stony Brook, NY   11794-3733

Dear Dr. Paldy:

I have enclosed a manuscript, "Teaching Technical
Writing to Undergraduate Chemistry Majors", which I am
submitting for publication in the Journal of College Science
Teaching.

I believe that this topic would be of interest to
members of the National Science Teachers Association (NSTA)
because it is applicable to all areas of the sciences,
although chemistry is used exclusively in the examples. The
writing abilities of students (potential employees) have
been of great concern to educators and employers, and perhaps the innovations mentioned in the paper can be used
successfully by others interested in this subject.

Three copies of the manuscript have been mailed to the
managing editor at NSTA Headquarters in Washington, D.C., as
directed in the journal's "Information to Authors". The
manuscript has not been published elsewhere and is not being
considered for publication by other journals.

I hope that you and the referees will find the topic
suitable for publication. I will be awaiting your response
and recommendations.

Sincerely yours,

B. Edward Cain

B. Edward Cain, Ph.D.
Professor of Chemistry

BEC:ec

Enclosure

Figure 7. Letter of transmittal for a journal article.

should be complete and sincere. It should not be condescending or excessive. State your apology simply, and indicate what accommodations or remedies you will make. Do not write unrealistic statements such as "I will make sure that this doesn't happen again", because you may not have complete control over whether it happens again. It would be better to write "I will make every effort to remedy this in the future."

If the apology involves a financial adjustment in the client's favor, mention this at the beginning of the letter. Everyone is happy to get this type of information, and the person to whom you are apologizing will be much more receptive to your other comments.

Letter of Congratulation

A successful professional will earn the respect of his or her associates by taking five minutes to write a letter of congratulation or appreciation when appropriate. The letter might be written to a college friend who has published recently or to a colleague who has retired. Depending upon your relationships with co-workers, you might extend best wishes upon a marriage, birth, graduation, or a child's success. The person receiving this short letter will be very pleased by your thoughtfulness.

Some situations, however, require restraint. A retirement is not always voluntary; a person may be retiring because of ill health or other unpleasant circumstances. If this is true, state the ways in which the person will be missed and wish her or him happiness in the future. Any personal touch that can be added, such as noting his or her fondness for travel or fishing or gardening, will be appreciated.

Letter of Recommendation

Let us assume that the person needing your recommendation has asked you to write in his or her support, and that you have been honest in stating your comfort or discomfort in doing so. The letter of recommendation should be very descriptive about the person's capabilities or potential that you are able to evaluate. Mention any deficiencies or needs for improvement as you assess the individual's capabilities. Many people make the mistake of writing letters filled with praise, and ignore any problems or special needs. These letters are seldom believed by the reader, who may question the integrity of the writer.

An uncritical letter can also cause problems for future applicants, because the reader may think you lack the ability to thoroughly assess a person's skills, and therefore not take your recommendations seriously.

Letter of Gratitude

A letter of gratitude (or a thank-you note) represents good manners both in business and social situations. A thank-you letter should be short and to the point. It should be personal and should mention the specific item(s) for which you are grateful. Your response should be as prompt as possible; delay only makes the letter more difficult to write.

Letter of Condolence

A letter of condolence is probably one of the most difficult letters to compose. Its purpose is to provide comfort to the reader; it should be compassionate while remaining direct and simple. A handwritten message is appropriate, but a typewritten letter is acceptable on personal stationery.

If the deceased was an employee, the letter may mention some of his or her contributions to the company or institution and how much he or she will be missed. A nice gesture would be to mention some of the qualities that made it pleasurable to have been a colleague. If the person endured a long, painful illness, mention his or her release from suffering, but be careful not to assume that the family is relieved by the death.

Unless you are very familiar with the family, avoid making religious comments; it would be considered rude to say that you are praying for the soul of someone who was an agnostic or atheist or whose religion did not include belief in a life after death. Also avoid making maudlin or excessively emotional statements.

The letter should be written and mailed on the day of notification. If you are offering assistance to the family, give specific details. When the letter is intended to represent the company, include a brief line that a representative of the company will contact the family soon about the benefits, personal effects, or insurance belonging to the deceased.

If a relative of an employee has died, a letter should be written to express your sympathy to the employee. The letter can be much shorter and less specific in details, depending upon your familiarity with the family. Sympathy cards are commercially available and abundant, but a letter has much more meaning. (Sending a card is better than sending nothing, however.)

Exercises

1. Write a letter of application responding to one of the advertisements in Figure 8.

2. Assume that you are a part-time employee, and write a letter that you would send to your employer if you wanted to become a full-time employee.

3. Compose a letter to a famous scientist requesting that he or she present a lecture for your local science professional group.

ANALYTICAL CHEMIST—BS

AB Chemical Company continues to research and develop quality personal-care products that are distributed nationally. Our Product Design Department can offer you rewarding new challenges and opportunities for your increased professional development.

Your responsibilities would include a wide range of activities, including product analysis and quality control, testing, and participation in research. Advancement to supervisory roles is possible.

To qualify, you must have a BS degree in Chemistry (or a related area) and experience in analytical chemistry. Abilities in technical writing, computer programming, and GC are a plus.

Compensation will be commensurate with experience. Please forward your résumé with salary requirements to Brenda Starkweather, Director of Product Design, AB Chemical Company, 444 Western Boulevard, Zenith, NY 14001–4502.

AB CHEMICAL COMPANY

An Equal Opportunity Employer

BIOCHEMIST OR BIOLOGIST

Immediate entry-level opening for BS-level Biochemist or Biologist at ZYX Fiterite Company, a leading producer of pesticides and insecticides. Opportunities for advancement are excellent.

Position requires laboratory skills in biochemical techniques involved with microanalytical chemistry. The successful applicant should have a BS degree in Biochemistry or Biology (or a closely related field). Experience in laboratory work is preferable.

Apply to: Mr. Leon Dixon, ZYX Fiterite Company, P.O. Box 709, Englewood Cliffs, NJ 07632 (An Equal Opportunity Employer)

Figure 8. Position advertisements.

| Chapter 17 |

Résumés

Are you seeking your first job? (Or looking for a new job?) Are you being considered for a promotion? Will you be giving a talk or a seminar? Have you been nominated for an award?

In any of these situations, an inventory of your professional qualifications, education, achievements, and experience will be required. This brief summary of your history is one of the most important documents of your professional life: your résumé.

The résumé highlights a person's educational and vocational background in a concise, but thorough, outline. A résumé is also called a curriculum vita (c.v., course of one's life; plural: curricula vitae) or a vita brevis (short life). A résumé is most commonly used during a job search, and the information presented in this chapter will be directed toward that use.

General Guidelines

Many acceptable formats can be used for presenting information in a résumé, and examples of style and content will be presented later in this chapter. Regardless of the style selected, some rigid guidelines must be followed.

✔ All of the information must be factual. You will naturally want to emphasize the high points of your achievements, background, and capabilities, but do not exaggerate information to the extent that it could be considered false. Remember that your history can be verified,

1451–4/88/0171$06.00/1 © 1988 American Chemical Society

and most employers check references or obtain copies of your transcripts and other official documents.

The other extreme should also be avoided. Many people seeking their first job downplay their abilities and experiences. Summer employment at a fast-food restaurant, although it may seem trivial, can impress a potential employer if a person's integrity, sense of responsibility, and ability to work with others were demonstrated while holding the position.

✔ The résumé must be typed or typeset. A busy employer or personnel director will simply not bother to consider a handwritten résumé as a serious application for employment. With the increasing availability of word processing equipment, complying with this requirement should not be difficult. (In larger cities, many businesses specialize in the preparation of résumés. Having your résumé typeset is often a worthwhile investment.) A new ribbon should be used to type a résumé. If a word processor is used, the output must be on a letter-quality printer. If photocopies are to be made, they must be of high quality—they should not appear to be photocopies!

✔ Grammar and spelling must be perfect. There is nothing worse than reading the résumé of an applicant who majored in "chemitry". If you cannot type an error-free copy, hire someone who can; it will be a small amount of money well spent.

✔ The résumé must be attractive. Remember that the person receiving your résumé has many others to consider, and you will want your résumé to make a positive impression. The résumé should not only provide information, but also arouse the potential employer's interest in you. A busy employer cannot and will not waste time trying to read light or illegible type, smudged ink, or an incoherently arranged presentation of information, no matter how impressive the candidate's credentials might be. Bond paper should be used; white was the only acceptable color in the past, but cream, beige, light blue, and gray are seen frequently. Occasionally, an applicant will have his or her professional photograph duplicated on the résumé, but gimmicks and drawings should be avoided.

✔ The résumé must be limited to one or two pages. This guideline can be violated occasionally, but keep in mind that the person reading the résumé may become impatient at having to scan through several pages to find pertinent information. Figures 1 and 2 show typical one-

---

BART J. SMITHERS

SCHOOL ADDRESS: PERMANENT ADDRESS:
Rochester Institute of Technology   658 Main Street
P.O. Box 9887   Akron, NY 14001
Rochester, NY 14623   (716) 542-0000
(716) 475-0000

PROFESSIONAL GOAL:
To use analytical chemistry skills in a quality control position.

EDUCATION:
Rochester Institute of Technology, Rochester, NY 14623
Attended   September 1984 to date
Degree   B.S. (anticipated June 1988)
Major   Chemistry
Major Subjects   Separations Techniques, Physical Chemistry, Advanced
                  Instrumental Analysis, Kinetics, Inorganic Chemistry, FORTRAN,
                  Calculus, Differential Equations, Physics, German
Grades   3.2 (A = 4.0)

SUNY Agricultural and Technical College, Alfred, NY 14802
Attended   September 1982 to June 1984
Degree   A.A.S., June 1984
Major   Chemical Technology
Major Subjects   General Chemistry, General Biology, Calculus, Quantitative
                  Analysis, Organic Chemistry, Instrumental Analysis, Introduction
                  to Physical Chemistry
Grades   3.3 (A = 4.0)

RELATED WORK EXPERIENCE:
November 1984   Eastman Kodak Company, Rochester, NY
to date   Cooperative work experience in Research Division; performed spec-
                  trophotometric analyses and chromatographic methods (HPLC, TLC,
                  and GC). Evaluations were "excellent". Worked 3 quarters, to-
                  talling 9 months.

OTHER WORK EXPERIENCE:
1982 to 1984   Summers. Batavia YMCA Camp, Batavia, NY.
                  Camp counselor and crafts director. Also responsible for running
                  camp store, maintaining inventory, ordering supplies, and book-
                  keeping.

REFERENCES:
Available upon request.

                                                                July 1987

Figure 1. Reverse chronological résumé.

SARA JUNE JOHNSON

47 Mustard Street, Apt. 4B
Rochester, NY 14609
(716) 271–0000

JOB OBJECTIVE: Entry-level management in research and development in areas of biotechnology or biochemistry.

EDUCATION:
Rice University, Houston, TX
27 semester credit hours of business and management courses
1984–1985

Rochester Institute of Technology, Rochester, NY
B.S., Biotechnology, June 1982
GPA 3.8 (A = 4.0)

SKILLS:
Grant proposals. Wrote successful grant proposal to NSF for $50,000 research project in biotechnology. Directed and coordinated work of three technicians.

Budgetary management. Responsible for ordering equipment, seeking bids, and authorizing purchase. Managed the maintenance of equipment within budgetary guidelines.

Laboratory skills. Devised new biotechnology techniques for bacterial strain development.

EMPLOYMENT:
July 1985 to present: BBB Biotechnology Labs, Rochester, NY
    Division manager, bacteria development.

July 1982 to July 1985: Texas Technology, Inc., Houston, TX
    Technologist, biotechnology department.

1979 to 1982: Student cooperative employment. Keele Company, Rochester, NY
    Technician, bacteriology and immunology.

PERSONAL DATA:
Single; excellent health. Active in local political organizations and community affairs. Prefer to remain in northeastern United States. Reading and writing abilities in French, German, and Russian.

REFERENCES:
Letters of recommendation will be sent separately.

June 1987

Figure 2. Résumé in the functional format.

page résumés, which will be discussed in more detail later in this chapter. Figure 3 shows an exception to this guideline. In this case, the person has a long publication and patent listing and a long, varied employment history.

✔   When possible, the résumé should be tailored to the specific job opportunity or situation. A person applying for his or her first professional job often uses mass mailings, and does not tailor the résumé for each employment opening. If a job description is available, try to address the particular requirements in your résumé or letter of application (see Chapter 16).

HOWARD W. MANNING

1234 Main Street
Lima, NY 14485
(716) 624–0000

CAREER GOAL: Upper-level administration in a Ph.D.-granting institution

EDUCATION:
Ph.D.   1950, Syracuse University, Syracuse, NY
          Major: Inorganic chemistry, X-ray crystallography
          Dissertation: "X-ray Structure Determination of Ammonium Sulfamate"

B.A.   1944, University of Buffalo, Buffalo, NY
          Major: Chemistry

EMPLOYMENT HISTORY:
1980–present: **Head and Professor**
   Department of Chemistry
   University of Rochester
   Rochester, NY 14627
      Duties include managing a department of 35 faculty and staff with an annual budget of several million dollars; supervising classroom instruction at the graduate and undergraduate levels; administering research grants; recruiting faculty and students; fund-raising for equipment and capital improvements.

1974–1980: **Professor**
1968–1974: **Associate Professor**
1962–1968: **Assistant Professor**
   Department of Chemistry
   Rochester Institute of Technology
   Rochester, NY 14623
      Duties included teaching and conducting research at the graduate (MS) and undergraduate levels; Acting Head (1979–1980); writing grant proposals; serving on committees and advisory groups.

Figure 3.   Résumé for an experienced worker. Continued on next page.

1956–1962: **Research Director**
1952–1956: **Research Associate**
  Photographic Emulsions Lab
  Eastman Kodak Company
  Rochester, NY 14650
    Duties included research and development; budget and personnel management;
    design of bench-scale processes.
1950–1952: Postdoctoral fellowship

PUBLICATIONS: A list of publications covering the most recent decade is attached.

PROPOSALS AND GRANTS: A list of funded proposals and grants covering the most
  recent decade is attached.

PRESENTATIONS: Professional presentations have been made at the local, state, and
  national levels and are too numerous to list. They will be provided upon request.

PROFESSIONAL MEMBERSHIPS:
  American Association for the Advancement of Science
  American Association of University Professors
  American Chemical Society
    Past-President, Rochester Chapter
    member, Chemical Education Division
    member, Inorganic Chemistry Division
  Sigma Xi

Attachments   MAY 1987

Figure 3.—Continued.

Contents of the Résumé

Although the format for a résumé can vary, the document should
contain all information pertinent to its purpose and should portray
an overall picture of you as a professional. The entries below are
usually necessary, but every person's background is unique, and the
contents of the résumé should be arranged to highlight your individuality.

1. Center your name, address, and phone number (including
   area code) at the top of the page. If your permanent or res-
   idential address is not the same as your mailing address, you
   may wish to include both. The date should appear (either at
   the top or the bottom of the page) to indicate that the information contained in the résumé is current. This fact is
   sometimes obvious (as with a recent or prospective graduate),
   but is important for someone who is already employed.

2. The next item might be your career objective(s) or goal(s). This brief statement should list short-term ambitions but can include long-range aims. If you tailor the career objective specifically to the job description, do so honestly. Geographical preference, if any, can be specified at this point.

3. A listing of work experience and educational background should follow the objectives. A person with no vocational experience in the specific professional area may decide to place the educational section before the list of jobs held, but a person with applicable job experience may wish to emphasize job experience by placing employment history before the educational information. The educational portion should contain the college(s) attended, degree(s) and date(s), major, minor (if appropriate), and courses that indicate your qualifications and training for the position. If you attended two or more colleges, list them in reverse chronological order. Similarly, if you received multiple degrees, the highest level should be the most prominent. A senior who is applying for a position before receiving a degree should indicate the date the degree will be received (as shown in Figure 1).

Any special training, not resulting in a degree, should be cited in this section. If you worked your way through college, include that information in this section or in the employment listing (see item 4 below). If you received awards or scholarships in college or participated in extracurricular activities, list these and other accomplishments here or emphasize them in a special section (see items 6 and 7 below).

Recent graduates often ask whether they must include their grade-point average (GPA). From a practical point of view, if your GPA is high, include it and indicate the basis of calculation (e.g., A = 4.0). If your GPA is low, omit it; the potential employer may see your transcript eventually, but there is no point in having a decision made on that basis so early in the process. Normally, high school background is not provided unless a high school diploma was the highest degree attained.

4. The employment section is sometimes divided into two parts: (1) experience related to your career or to the job for which you are applying, and (2) experience in other jobs. With either category, list the positions in reverse chronological order. Include the name and address of the employer, your job title, and a brief description of the position, emphasizing skills, achievements, and specific duties and responsibilities that are applicable to the position you seek. Use action verbs to de-

scribe your responsibilities; for example, administer, analyze, create, delegate, develop, direct, establish, evaluate, implement, improve, increase, initiate, maintain, manage, organize, originate, outsell, recruit, represent, supervise, train.

Mention work experience unrelated to your profession but describe it in less detail. Part-time jobs held while in college can be included, especially if you have graduated recently. Part-time positions indicate that you have the ability and desire to obtain and keep a job. Any signs of initiative and perseverance will be looked upon favorably by the employer.

5. Experience in the Armed Forces should be listed if applicable. This information is essential if gaps in the educational or vocational history can be explained by including dates of military or naval service. It is appropriate to include rank and promotions, specific training, leadership experience, discharge status, and skills obtained in the service. Include this information in the education or employment sections mentioned previously.

6. Activities and personal interests are often included in the résumé. If you hold memberships in professional organizations (or student affiliate groups), they should be mentioned. Involvement in community or extracurricular activities, hobbies, and other interests can be inserted here.

7. Honors, awards, or scholarships show exemplary performance and provide evidence of your character and abilities as judged by others. As mentioned previously, these accomplishments could be listed in the educational section or separately to provide emphasis.

8. Most sample résumés include a section containing personal data: age, weight, height, marital status, and health. Employers usually do not pay much attention to this information; furthermore, they are required by antidiscrimination laws not to be influenced by these statistics. Some people debate whether the category should be inserted at all, because a prospective employer may unconsciously assimilate this information. Most personnel directors advise you to include this information only if you are comfortable doing so and if you think it will be beneficial.

People with permanent physical disabilities, such as visual, hearing, or mobility impairment, often question whether a physical disability should be mentioned on the résumé. At times the condition is obvious from other information that appears on the résumé (e.g., if the applicant received a degree from the National Technical Institute for the Deaf or held membership in the American Association for the Blind). In most instances, the cover letter, rather than the résumé, is the best place to indicate special situations. The letter can explain why the physical disability would not interfere with job performance and any allowances that would be necessary (e.g., wheelchair access, guide dogs, interpreters). Many scientists who have physical disabilities feel that the job interview is the appropriate place to discuss these matters.

9. References should be the final entry on the résumé. Stating that references will be supplied upon request is often sufficient. (This practice is especially convenient if the résumé is being printed in large quantities and will be used for a variety of jobs.) You may use different people as references for different types of positions, and it may be easier to give their names in the cover letter. If the job advertisement specifies that references be included with the application, they should be listed here (always with permission of the persons used as references). Include phone numbers if they are available, because many interviewers prefer to phone references to get a more personal interpretation of the applicant's abilities. If a college or professional placement service is handling your documentation, provide the name and address of the service.

The items just mentioned should be included in almost every résumé. Because the résumé reflects you as an individual, you may wish to provide pertinent information that has not been mentioned in other categories, such as travel experience, proficiency in foreign languages, research interests, and other important areas of your life.

Formats for Résumés

As mentioned, your résumé should reflect you as a professional and it should, if possible, be designed for a specific position or purpose. The two most popular ways to present this information are the reverse chronological format and the functional format.

Reverse Chronological

Reverse chronological is the most commonly used form. Educational and vocational experiences are listed with the most recent first. Because this format emphasizes positions and organizations, it would be used by people who want to highlight their last employer, an important job title, or a well-known company. A person who wants to advance in the same field might use this style, but if he or she is making a radical change in careers, the alternative functional format may be more advantageous.

A recent graduate will probably find the chronological style more effective because it emphasizes educational preparation for employment. An applicant who wants to minimize the focus on age (old or young) should avoid this style because it stresses dates. An example of a chronologically arranged format is shown in Figure 1.

Functional Format

The functional format emphasizes skills that are applicable to the position sought. These attributes might include management skills, dealing with the public, communication abilities, fund-raising successes, or skills demonstrated in volunteer and community activities or through compensated employment.

Who should consider using this format? If a person is changing careers or has had several unrelated positions, job titles are less important than duties or responsibilities. This condition will be true for people who have had interruptions in their careers. If your employment history has centered around the same type of activity or has demonstrated growth within an institution or position, the functional approach will not be as effective as the chronological approach. Figure 2 provides a typical example of the functional résumé.

Discrimination Through a Résumé

Almost every advertisement for a professional position states that the employer is an Affirmative Action/Equal Opportunity Employer, meaning that the institution hires men and women, veterans, and disabled individuals of any race, color, and national or ethnic origin. In some areas of the country, marital status and sexual preference cannot be used as a basis for selection (or rejection) of applicants.

As mentioned previously in this chapter, the typical résumé can provide information that could enable an employer to discriminate. Indeed, almost any information given could be regarded as a potential basis for discrimination. A person's name usually denotes gender, although gender can be obscured temporarily by using initials (e.g., M. A. Jones could be Mary Ann or Martin Alan). A surname could provide clues to ethnic origins, race, or religion. If the employer thinks the college attended by an applicant is second-rate or less prestigious than other colleges, such information can prejudice an employer against an applicant.

How cautious should you be, therefore, without becoming paranoid? Again, the only answer is that you should give information with which you feel comfortable, while remaining within the guidelines of honest representation of your experience and background. Fortunately, antidiscrimination laws and the good intentions of most employers have minimized the amount of discrimination that occurred in the past.

Exercises

1. State your short-term career objective(s) in one sentence.

2. List the skills (that you already possess) that would help you to achieve your career objective(s). What skills are you lacking?

3. Prepare your current résumé using (a) the reverse chronological format and (b) the functional format.

4. Using one of the biographical reference works, such as *Who's Who in America* or *World Who's Who in Science* (both published by Marquis Who's Who Inc., Chicago, IL), choose a well-known living professional in your area of expertise and prepare a reverse chronological résumé based on his or her life.

| Chapter 18 |

Memos and Short Reports

The office memorandum (plural: memorandums or memoranda) or memo is unavoidable if you work in an industrial, research, or academic setting. You will be receiving them, writing them, or both depending upon the nature of your position. Although memos are used for communicating within a place of employment, and business letters (*see* Chapter 16) are normally used for external situations, they share many of the same functions.

- Memos can be used to request information or services from another person or group within the organization. A supervisor could ask for a cost-comparison estimate before purchasing new equipment; a colleague might request information for a special project; an employee might use a memo as a cover letter for a proposal requesting that the employer consider an idea or suggestion for improving work conditions.

- A memo might be written to share or provide information of all types—from policy matters and technical reports to the time of the next office party or bowling league banquet. Opinions, suggestions, and recommendations can be solicited or given by memo. Progress reports, cost estimates, instructions for equipment use, and descriptions of problem areas all are examples of providing information.

- Another increasing function of a memo is to document or record an activity such as a meeting or phone call. Notification of a meeting time and location and the minutes of that meeting

1451–4/88/0183$06.00/1 © 1988 American Chemical Society

might both be in memo format. When used in this way, the memo is filed as a permanent or semipermanent record. An employee who does not agree with a recent decision might prepare a memo outlining objections for the purpose of having the objections on the record. Memos have been used as evidence in litigation proceedings; this fact emphasizes the importance that memos be written precisely and accurately, as is true for all technical writing.

- Finally, a memo can function as a reporting tool. It might be used to account for activities, such as a site inspection, business trip, or research results. A marketing report for a proposed new product could be presented in memo form as could a compilation of survey results.

Very often, it is difficult to categorize a memo under one of the areas just mentioned, because a tremendous amount of overlap exists. A proposal in the form of a memo, for example, would be likely to contain a request, provide information, report on previous findings, and act as a record of existing situations and recommendations.

Format of Memos

Most large organizations use a standard form for office memos. Typically, the organization's name and logo appear at the top, and spaces are provided for the name(s) of the person(s) receiving the memo, the name of the person writing the memo, the date, and the title or subject of the memo.

A memo is usually typed, although it can be handwritten if the occasion and purpose of the memo are rather informal. Unless your handwriting is very legible, type the memo. If no printed form exists, the same information can be typed:

MEMORANDUM

TO:

FROM:

DATE:

SUBJECT:

The title and first sentence of a memo are very important. Because people receive a large number of memos, some memos are immediately filed (often in the circular file) without being read if the preliminary information appears to be of no concern. In a large organization, filing clerks may need to use the information given as the subject to decide how the memo should be filed. Figure 1 gives an example of a meeting notification memo.

If a memo will be sent to a large number of people who need to be specified by name, the distribution is placed at the end of the memo with a notation in the heading, as shown in Figure 2.

MEMORANDUM

TO: C. Jones, R. Smith, M. Maple, J. Johnson

FROM: E. Cain

DATE: April 1, 1987

SUBJECT: CHEMISTRY CURRICULUM COMMITTEE
               MEETING, APRIL 3

The next meeting of the departmental Curriculum Committee will be at 1:00 p.m. on Friday, April 3, in the Dean's Conference Room. The agenda will consist of the following:

1. course approval for Inorganic Chemistry revisions (copies were distributed at the last meeting)

2. consideration of a request from the School of Printing for a change in the courses required for Printing majors (see attached memo)

3. miscellaneous business

Attachment

cc: T. Potter, Head, Department of Chemistry

BEC:ec

Figure 1. A meeting notification memo.

```
                              MEMORANDUM


    TO:              Distribution below*

    FROM:            E. Cain

    DATE:            March 27, 1987

    SUBJECT:         SCHEDULING TIME FOR USE OF NMR


    (Text of the memo)


    *DISTRIBUTION:   Person 1
                     Person 2
                     Person 3
                     Person 4, and so on.
```

Figure 2. A memo listing distribution.

Depending upon the formality of the memo, titles can be appropriately used in the heading:

MEMORANDUM

TO:	Dr. C. Lyon, Head, Department of Zoology
FROM:	Dr. Ed Cain, Chair, Chemistry Curriculum Committee
DATE:	February 28, 1987
SUBJECT:	CHEMISTRY COURSES FOR ZOOLOGY MAJORS

To determine the length of a memo, follow this rule: the shorter, the better. However, a long memo is preferable to one that has incomplete information. Make your message clear and readable. This effort is often ignored because people know each other well within a company and tend to become sloppy. As with all aspects of technical writing, the memo should be free of errors. I recently received a memo that had so many typographical errors and grammatical mistakes that it was nearly impossible to understand what was intended, and because it was generated on a word processor, excuses about not being a good typist would not hold.

If the memo exceeds one page, succeeding page(s) should have headers with the recipient's name, the date, and the page number. Alternatively, the memo subject could replace the recipient's name, especially if distribution is to a large group of people.

Unlike a letter, a memo has no closing other than a summary sentence. If a signature is required, either the initials or signature of the sender are placed near the name in the memo heading. At the end of the memo, the typist's initials, distribution of copies, and attachments should be noted. ("Enclosure" is normally reserved for letters, and "Attachments" is more appropriate for materials accompanying memos.)

Exercises

1. Write a memo to the head of your department or chairperson about one of the following:

   (a) a suggestion for improvement(s) within the department
   (b) a request for funds for a student activity
   (c) documentation of a problem you are having with your major, one of your professors, or your advisor
   (d) a request for information about graduate schools

2. Write a memo that informs your supervisor, professor, or club president about the progress you are making on a special project.

INDEX